Ponnadurai Ramasami (Ed.)
Computational Sciences

Also of interest

Computational Strong-Field Quantum Dynamics
Intense Light-Matter Interactions
Bauer (Ed.), 2017
ISBN 978-3-11-041725-8, e-ISBN 978-3-11-041726-5

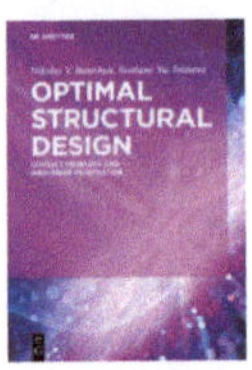

Optimal Structural Design
Contact Problems and High-Speed Penetration
Banichuk, Ivanova, 2017
ISBN 978-3-11-053080-3, e-ISBN 978-3-11-053118-3

Computational Physics
With Worked Out Examples in FORTRAN and MATLAB
Bestehorn, 2018
ISBN 978-3-11-051513-8, e-ISBN (PDF) 978-3-11-051514-5

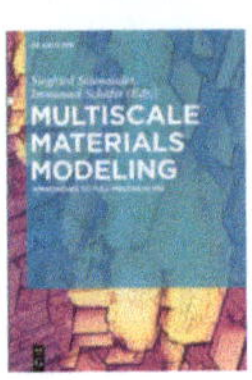

Multiscale Materials Modeling
Approaches to Full Multiscaling
Schmauder, Schäfer (Eds.), 2016
ISBN 978-3-11-041236-9, e-ISBN (PDF) 978-3-11-041245-1

Computational Sciences

Edited by
Ponnadurai Ramasami

DE GRUYTER

Editor
Prof. Ponnadurai Ramasami
University of Mauritius
Department of Chemistry
Réduit 80837, Mauritius
p.ramasami@uom.ac.mu

ISBN 978-3-11-046536-5
e-ISBN (PDF) 978-3-11-046721-5
e-ISBN (EPUB) 978-3-11-046548-8

The articles in this book have been previously published in the journal *Physical Sciences Reviews* (ISSN 2365–659X).

Library of Congress Cataloging-in-Publication Data
A CIP catalog record for this book has been applied for at the Library of Congress.

Bibliographic information published by the Deutsche Nationalbibliothek
The Deutsche Nationalbibliothek lists this publication in the Deutsche Nationalbibliografie; detailed bibliographic data are available on the Internet at http://dnb.dnb.de.

Cover image: KTSDESIGN/Science Photo Library/getty images
Typesetting: Integra Software Services Pvt. Ltd.
Printing and binding: CPI books GmbH, Leck
♾ Printed on acid-free paper
Printed in Germany

www.degruyter.com

Preface of the Book of Proceedings of the Virtual Conference on Computational Science (VCCS-2016)

A virtual conference on computational science (VCCS-2016) was organized online from 1st to 31st August 2016. This was the fourth virtual conference which was started in 2013. The month of August was chosen to commemorate the birth anniversary of Erwin *Schrödinger*, the father of quantum mechanics, on 12th August.

There were 30 presentations for the virtual conference with 100 participants from 20 countries. A secured platform was used for virtual interactions of the participants. After the virtual conference, there was a call for full papers to be considered for publication in the conference proceedings. Manuscripts were received and they were processed and reviewed as per the policy of *De Gruyter*.

This book is a collection of the eleven accepted manuscripts. These manuscripts cover a range of topics from fundamental to applied science using computational methods. Marcano investigated anthocyanidin and anthocyanin pigments for dye sensitized solar cells based on Density Functional Theory (DFT) method. DFT method was also used by Ranjan *et al.* to study silver-gold nanoclusters. Gbayo *et al.* applied DFT method to understand the mechanism of nucleophilic substitution reactions of 4-(4'-nitro)phenylnitrobenzofurazan ether with aniline in acetonitrile. Khairat *et al.* used high level ab initio method to probe the electronic states of two newly detected dications namely MgS^{2+} and SiN^{2+}. Palafox applied several *ab initio* methods to vibrational spectroscopy for a better characterization and assignment of all the bands of the experimental spectra. Majee and Biswas used computational algorithm methods where Spatial Aggregation Propensity was employed and molecular dynamics simulation approach for prediction of aggregation prone areas in monoclonal antibodies. Chowdhury *et al.* studied the optical and magnetic properties of free standing two-dimensional (2D) materials silicene, germanene and T-graphene while the contribution by Schwingenschlögl *et al.* provides a review of elemental two-dimensional materials beyond graphene. Ocaya and Terblans addressed the challenges of standalone multi-core simulations in molecular dynamics. Patade and Bhalekar analyzed pantograph equation with incommensurate delay and provide analytical solution. Chooramun *et al.* used the Hybrid Space Discretisation to simulate evacuation and this was applied to large underground rail tunnel station.

I hope that these chapters will add to literature and they will be useful references.

To conclude, VCCS-2016 was a successful event and I would like to thank all those who have contributed. I would also like to thank the Organising and International Advisory committee members, the participants and the reviewers.

We are currently planning for the VCCS-2017 to be held from 1st to 31st August 2017.

https://doi.org/10.1515/9783110467215-202

Prof. Ponnadurai Ramasami
Computational Chemistry Group, Department of Chemistry, Faculty of Science, University of Mauritius, Réduit 80837, Mauritius
E-mail address: p.ramasami@uom.ac.mu

Contents

List of contributing authors

Chapter 1
Dr. R.O. Ocaya
University of the Free State – Qwaqwa Campus
Physics
P. Bag X13
9866 Phuthaditjhaba, South Africa
E-mail: ocayaro@ufs.ac.za

Prof. Koos Terblans
University of the Free State
Physics
Bloemfontein
South Africa
E-mail: terblansjj@ufs.ac.za

Chapter 2
Dr. Jana Debnarayan
University of Calcutta
Physics
92 A P C Road
700073 Kolkata, India
E-mail: cujanad@yahoo.com

Prof. Namrata Dhar
University of Calcutta
Department of Physics
Kolkata
West Bengal, India
E-mail: namrata.dhar05@gmail.com

Prof. Arka Bandyopadhyay
University of Calcutta
Department of Physics
Kolkata
West Bengal
India
E-mail: arkbndy@gmail.com

Dr. Suman Chowdhury
University of Calcutta
Department of Physics
92 A.P.C. Road
700009 Kolkata
West Bengal, India
E-mail: sumanchowdhury88@gmail.com

Chapter 3
Prof. Najia Komiha
Universite Mohammed V Agdal
of chemistry
Av.Ibn Batouta
Agdal
Faculty of sciences
10000 Rabat, Morocco
E-mail: komiha@fsr.ac.ma

Dr. Khadija Marakchi
University Mohammed V-Rabat-Morocco
LS3ME
Rabat, Morocco
E-mail: marakchi_k@yahoo.fr

Mohammed Salah
University Mohammed V-Rabat-Morocco
LS3ME
FS Rabat
1002 Rabat, Morocco
E-mail: salahmed53@hotmail.fr

Toufik Khairat
University Mohammed V-Rabat-Morocco
LS3ME
Rabat, Morocco
E-mail: marakchi_k@yahoo.fr

Chapter 4
Dr. Sachin Bhalekar
E-mail: sachin.math@yahoo.co.in

Dr. Jayvant Patade
Department of Mathematics
Shivaji University
Kolhapur
416004 Maharashtra, India
E-mail: jayvantpatade1195@gmail.com

Chapter 5
Mauricio Alcolea Palafox
Universidad Complutense de Madrid
Physical Chemistry
ciudad Universitaria

28040 Madrid, Spain
E-mail: alcolea@quim.ucm.es

Chapter 6

Dr. K. Lobb
Rhodes University
Department of Chemistry
Chemical and Pharmaceutical Sciences
Cnr University/Artillery Roads
6140 Grahamstown, South Africa
E-mail: k.lobb@ru.ac.za

K. Gbayo
University of Lagos
Department of Chemistry
Akoka, Yaba
23401 Lagos, Nigeria
E-mail: kgbayo@unilag.edu.ng

Dr. C. Isanbor
University of Lagos
Department of Chemistry
Akoka, Yaba
23401 Lagos, Nigeria
E-mail: cisanbor@unilag.edu.ng

Dr. O. Oloba-Whenu
University of Lagos
Department of Chemistry
Akoka, Yaba
23401 Lagos, Nigeria
E-mail: ooloba-whenu@unilag.edu.ng

Chapter 7

Dr. Sutapa Biswas Majee
NSHM College of Pharmaceutical Technology
Division of Pharmaceutics
NSHM Knowledge Campus, Kolkata-Group of Institutions
124 B.L. Saha Road
700053 Kolkata, India
E-mail: sutapabiswas2001@yahoo.co.in

Chapter 8

Prof. Tanmoy Chakraborty
E-mail: tanmoy.chakraborty@jaipur.manipal.edu

Chapter 9

Dr. Nitish Chooramun
University of Mauritius
Software and Information Systems
Reduit, Mauritius
E-mail: n.chooramun@uom.ac.mu

Dr. Peter Lawrence
University of Mauritius
Computer Science and Engineering Department
Reduit, Mauritius
E-mail: P.J.Lawrence@gre.ac.uk

Prof. E. R. Galea
University of Greenwich
Fire Safety Engineering Group
SE10 9LS London, UK
E-mail: E.R.Galea@gre.ac.uk

Chapter 10

Dr. Emildo Marcano
Universidad Pedagogica Experimental Libertador
Instituto Pedagógico de Caracas, Centro de Investigaciones en Ciencias Naturales (CICNAT)
1030 Caracas, Venezuela
E-mail: jedi148@gmail.com

Chapter 11

Udo Schwingenschlögl
Physical Science and Engineering Division (PSE)
King Abdullah University of Science and Technology (KAUST)
Thuwal 23955-6900
Saudi Arabia
E-mail: udo.schwingenschlogl@kaust.edu.sa

Jiajie Zhu
Physical Science and Engineering Division (PSE)
King Abdullah University of Science and Technology (KAUST)
Thuwal 23955-6900
Saudi Arabia

Tetsuya Morishita
Research Center for Computational Design of Advanced Functional Materials

National Institute of Advanced Industrial Science and Technology (AIST)
Tokyo, Japan

Michelle J. S. Spencer
School of Science
RMIT University
GPO Box 2476
Melbourne
Victoria 3001
Australia

Paola De Padova
Consiglio Nazionale delle Ricerche - ISM
Via del Fosso del Cavaliere 100
00133 Roma
Italy

Aix-Marseille University
IMéRA
Marseille
France

Amanda Generosi
Consiglio Nazionale delle Ricerche - ISM
Via del Fosso del Cavaliere 100
00133 Roma
Italy

Barbara Paci
Consiglio Nazionale delle Ricerche - ISM
Via del Fosso del Cavaliere 100
00133 Roma
Italy

Carlo Ottaviani
Consiglio Nazionale delle Ricerche - ISM
Via del Fosso del Cavaliere 100
00133 Roma
Italy

Claudio Quaresima
Consiglio Nazionale delle Ricerche - ISM
Via del Fosso del Cavaliere 100
00133 Roma
Italy

Bruno Olivieri
Consiglio Nazionale delle Ricerche - ISAC
Via del Fosso del Cavaliere 100
00133 Roma
Italy

Eric Salomon
Aix-Marseille University
CNRS
PIIM
UMR 7345
Marseille
France

Thierry Angot
Aix-Marseille University
CNRS
PIIM
UMR 7345
Marseille
France

Guy Le Lay
Aix-Marseille University
CNRS
PIIM
UMR 7345
Marseille
France

Harold J. W. Zandvliet
Physics of Interfaces and Nanomaterials and
MESA+ Institute for Nanotechnology
University of Twente
P. O. Box 217
7500AE Enschede
The Netherlands

L. C. Lew Yan Voon
University of West Georgia
Carrollton
GA 30118
USA

R.O. Ocaya and J.J. Terblans

1 Addressing the challenges of standalone multi-core simulations in molecular dynamics

Abstract: Computational modelling in material science involves mathematical abstractions of force fields between particles with the aim to postulate, develop and understand materials by simulation. The aggregated pairwise interactions of the material's particles lead to a deduction of its macroscopic behaviours. For practically meaningful macroscopic scales, a large amount of data are generated, leading to vast execution times. Simulation times of hours, days or weeks for moderately sized problems are not uncommon. The reduction of simulation times, improved result accuracy and the associated software and hardware engineering challenges are the main motivations for many of the ongoing researches in the computational sciences. This contribution is concerned mainly with simulations that can be done on a "standalone" computer based on Message Passing Interfaces (MPI), parallel code running on hardware platforms with wide specifications, such as single/multi- processor, multi-core machines with minimal reconfiguration for upward scaling of computational power. The widely available, documented and standardized MPI library provides this functionality through the MPI_Comm_size (), MPI_Comm_rank () and MPI_Reduce () functions. A survey of the literature shows that relatively little is written with respect to the efficient extraction of the inherent computational power in a cluster. In this work, we discuss the main avenues available to tap into this extra power without compromising computational accuracy. We also present methods to overcome the high inertia encountered in single-node-based computational molecular dynamics. We begin by surveying the current state of the art and discuss what it takes to achieve parallelism, efficiency and enhanced computational accuracy through program threads and message passing interfaces. Several code illustrations are given. The pros and cons of writing raw code as opposed to using heuristic, third-party code are also discussed. The growing trend towards graphical processor units and virtual computing clouds for high-performance computing is also discussed. Finally, we present the comparative results of vacancy formation energy calculations using our own parallelized standalone code called Verlet–Stormer velocity (VSV) operating on 30,000 copper atoms. The code is based on the Sutton–Chen implementation of the Finnis–Sinclair pairwise embedded atom potential. A link to the code is also given.

Keywords: molecular dynamics, threads, MPI, standalone computation

https://doi.org/10.1515/9783110467215-001

1.1 Introduction

The last few decades have seen a rapid rise of computational modelling as a powerful tool to develop materials by postulation, inference and tuning in what can only be described as a semi-closed loop approach. The advantage of this approach is that properties that may not be directly ascertained, such as homogeneous nucleation [1], or are costly to investigate empirically can be evaluated [2–6]. A molecular simulation begins with choosing the best force field that describes the molecular configuration and ends with the iterative calculation of the phase space under specified boundary conditions that denote the external influences. Such conditions convey the applied forces, thermostats [7], pressure, etc. The choice of mathematical model, which in effect denotes the force field, affects the simulation data structures and therefore the design of the software. This in turn affects what can ultimately be parallelized. In our own work, we focus on the Sutton–Chen implementation of the Finnis–Sinclair embedded atom model potential [3, 7–9]. Force fields describe the interactions of particles in the configuration in terms of normalized dimensionless perturbations. The general assumption is that the internal force field in the molecular configuration is usually much stronger in comparison with the external perturbation. The required phase can be determined deterministically by molecular dynamics (MD), or stochastically through Monte-Carlo approaches [7, 10]. Ultimately, the results of such computation must be compared with the experiment to ascertain its accuracy or even to gauge the efficacy of the model. This leads either to a tuning of the model if at the model development stage or to a tuning of the material properties by parameter adjustment if the model has been tested and is accepted previously. There are generally two simulation approaches used, namely MD and ab-initio methods [11–20]. The basic computational modelling approach takes the aggregated, macroscopic properties arising as the consequence of their isolated, pairwise and interactive atomistic dynamics and considering the cumulative effects of these atomistic interactions as aggregations by some method that incrementally integrates their equations of motion. The recurrent idea in both approaches is to first derive a mathematical model that best describes the system, through which a state trajectory can be found. The mathematical model and its effectiveness and range depend on a number of factors that ultimately decide the extent of success of the simulation. Therefore, for a given system, it is imperative to derive a model that is as realistic as possible. However, this particular goal must take into consideration the level of accuracy required and the computational resources available. For atomic and molecular systems of macroscopic significance, the number of particles in the system can be extremely large, and studying their pairwise interactions will generate vast amounts of data and consume equally vast amounts of system time and resources. Using a faster processor does not necessarily alleviate the problem immediately, in spite of any mathematical simplifications such as specification of a cut-off distance for the interactions, simplification of the force field and so on that may have been devised. As interest in the computational study of materials grew, the issue of computational speed and accuracy quickly took centre stage and highlighted the shortcomings of the typical computer for these purposes. Compounding the situation is the fact that conformity to Moore's law [21, 22] is no longer assured using the device interconnect-size concept on which the law is based. Moore's law predicts a doubling of computational power approximately every 2 years as a consequence of the falling distances of the component interconnections on a semiconductor wafer. In 1971, the interconnect distance was around 10 μm and currently stands around 10 nm. It is expected to reach 5 nm around 2020. This distance affects the fastest speed that a charge conveying bit information can travel between any two devices on the same chip. Operating speed, however, is not simply a consequence of the interconnect distance. Other engineering considerations must be made, such as how heat dissipation will be handled. Also, as the physics shows, this distance is the realm of the lattice parameter (around 10 times the 0.361 nm lattice parameter of copper, for instance) where quantum mechanical effects manifest significantly. Consequently, processor speeds have more or less stabilized over the last 5 years at clocking frequencies of around 3 GHz. Clearly, to keep up with the demands of increasing computational power, there had to be a shift in the thinking. Today, vast increases in computational power are therefore not so much due to unilateral improvements in processor technology, but due to ingeniously interconnected multi-core, multiprocessor arrays. The development of computational frameworks and efficient algorithms appears to be driving innovations in atomistic MD simulations [7–9]. This is unwittingly a consequence of advancements in computer gaming, where the demand for high-performance computer graphics for more realistic rendering of ever larger and often networked games is ever greater and far outweighs other applications in the public domain. Thus, better models and computational algorithms are increasingly possible that exploit these features and are supported by ongoing developments in internet and computer technology (ICT). The rise of high-performance, supercomputing clusters [6] as low-cost supercomputers may perhaps be considered a secondary revolution in computation. This is because micron-scale, microprocessor interconnects [23] are now falling as fast as the demands of processing power require. However, in spite of these trends, standalone computation is still of considerable importance because it underscores the role of the individual node computer which is an integral component of any cluster, particularly if its inherent power could be unlocked through careful software design. The latter remains the basic computational element, and several such distributed nodes collectively execute chunks of tasks assigned by a task manager, which is also responsible for collating the results of the processing and making it available to the requester. Unfortunately, apart from the nature of interconnects between processors or cores themselves, the design of the node can easily become the weakest link in a computational arrangement and can lead to overheads whose cumulative effects substantially diminish the overall system performance. By standalone, one does not mean that the system is necessarily single processor or single core, but only that the computing node does not feature as part of a distributed system. Standalone can therefore mean any arrangement on a physically isolated, single machine that can support anywhere from single-core to multiple single-core processors to multiple processor multi-core computers. Such arrangements are today quite ubiquitous as standard features on desktop computers. For instance, the i3, i5 and i7 classes of x86-based machines common on desktops today are single processors with two to four cores and can support hyper-threading, such as Turbo Boost and K-technology [24]. The K-technology allows the overclocking of the CPU but for computational purposes illustrates the law of diminishing returns since it can limit the CPU power to only one core. Only applications written with one core in mind will likely reap any speedup benefit from such overclocking. The i7-Extreme can have up to eight cores and offer a unique processing opportunity if properly coded for the extra capability. Typically, most desktop applications do not use that extra feature and therefore do not observe the extra speedup. Additionally, on the same main board supporting the processor, there can be a high-definition hardware graphics processing unit (GPU). This multi-core CPU/GPU combination is the "typical" standalone configuration being referred to here. In our own work, the simulations are done on a Dell Optiplex 3010, which hosts a third-generation i5-3470 (3.2 GHz) with four cores and L3 cache (6 MB), has no hyper-threading but has K-technology (not activated to the maximum 3.6 GHz in this work). The system comes preinstalled with 8 GB (DDR3–1.6 GHz) of SDRAM. The GPU is AMD Radeon HD 7470 with 1 GB DDR3 SDRAM and interfaces the CPU with a PCI Express, 16-bit bus. Our results presented below are based on this hardware specification. The external storage, which is not part of the benchmark performance and mentioned here for completeness, is a 500 GB hard drive. Its function is merely to hold initial particle data and boundary conditions and to store the ultimate output of a simulation. Once the simulations start, no intermediate file access is done, all intermediate calculations being done in a SDRAM scratch pad. In the typical computer, the random access memory (RAM) is shared between cores or processors through a common, managed memory area called a "heap." There have also been dramatic developments in physical storage devices associated with the modern desktop to address the demands of traditional multimedia data files, if not for the more demanding storage of large gaming applications and data. For computational applications, physical storage media may be used as swap space and intermediate storage, although in the interest of speed they are generally relegated to input and ultimate data output. These considerations put a heavy demand, and therefore stipulation of type and layout, on the RAM requirements of such a system for successful computational modelling application. As a benefit of falling costs, computing clusters are becoming more common in medium-sized research facilities and universities. However, they are still generally still beyond the reach of many smaller organizations and individual researchers for a number of reasons. These reasons may revolve around an inability to justify their costs by any sort of sustainable research output, external third-party usage, running costs such as steady electrical power requirements which are likely to be proportional to the size of the system, availability of skilled manpower for configuration and maintenance, programming expertise for task scheduling, interpretation and result integration and so on. The starting inertia in computational research therefore tends to be high for the same reasons. As the popularity of computation continues to rise, the bulk of research output remains largely experimental, leaving computation largely unexplored.

As mentioned above, large MD simulations are today being run regularly for large particle systems with ever-increasing spatiotemporal scales. New innovations or extensions to existing systems are constantly being devised for increased computational performance. Many of these developments revolve around identification of code aspects that can be further parallelized or optimized. Parallelization starts typically with a decomposition, where the problem is divided into smaller “chunks” which are then assigned to a specific processor or core. There are three decomposition methods that are widely used to parallelize MD simulations. These are particle, force and domain decomposition using a spatial basis [25–27]. The particle number tends to be large in practically meaningful simulations, and none of these decompositions is yet very effective, due to memory consumption, a large number of processors and communication partners. Spatial decomposition approaches, based on physical subdomains, are generally superior [28]. In recent years, under the technological limitations imposed by Moore’s law, increases in computational power were achieved in two ways, namely increasing the number of cores in a processor and the vector size in single-instruction, multiple data instruction set (SIMD) computers. It has become almost *de facto* in this quest that the only avenue available to increase computational power is by an upward scaling of the number of processors and cores per processor in a chip. For a given configuration of processors, other performance enhancements can be devised though improved algorithms. It is clear that tapping into the full power of such a system will require a highly scalable parallel code. Unfortunately, the secondary issue of power dissipation and age degradation failure (e. g. due to any of the billions of transistors failing because of heat subjection) has to be considered carefully [29, 30]. Domain decomposition can present problems for inhomogeneous systems. The cell-task method [31] can overcome some problems, particularly if used as part of a hybrid decomposition scheme. For instance, by the use of a hybrid algorithm, Mangiardi and Meyer [29] devise a domain decomposition and thread-based hybrid parallelization technique based on large vectorization SIMD processors such as the AVX, AVX-2 and Xeon-Phi processors, comprising several thousands of cores. Their hybrid technique involves modelling with short-ranged forces for both homogeneous and inhomogeneous collections consisting of up to tens of millions of atoms. The hybrid algorithm achieves parallelization by using a task-based approach on smaller sub-domains obtained from domain decomposition. Each sub-domain is handled by a team or “pool” of threads running on multiple cores. The flow of execution of tasks cannot access the same particle simultaneously, thereby averting the need for lockout code to prevent such access of a particle by different tasks. At present, several parallelized MD programs are in existence, for instance [32–36] and others. However, such programs employ rigid parallelization techniques, and completely new versions will be needed to implement improved parallel algorithms that better exploit advancements in computer architectures [29]. In light of this proliferation, a smaller code that is easier to revise with algorithmic and architectural advancements has clear advantages, and is arguably necessary if advances in architectures are to be exploited fully in future [29, 37]. The effort required for such revisions can be small. There are some issues with hybrid methods, such as synchronization of data structure access which can lead to false sharing of these records because all processors constantly amend the data structures of neighbour cells [38]. These irregular memory access patterns pose significant challenges for parallelization and performance optimization. Additional challenges are associated with the hardware limitations that effectively limit active core performance due to chip configuration, such as memory speed, volume, interchip communication bandwidths, power management, etc. In 2010, Peng et al. [25–27] proposed three incremental optimizations for an emergent platform, the Godson-T multi-core architecture [39], which was formally released in late 2010. They presented behavioural, instruction-level simulations of a chip that was still in development and aimed at petaflops supercomputing. Their work addressed the foregoing issues of power efficiency, performance, programming and parallelism and platform-operating system portability (particularly to Linux and BSD; portability to WindowsCE is reported) by exploiting polymorphic parallelism. Its main contributions to high-performance computing (HPC) is improved interconnection data communications (both internal and external), fine-grained division of threads, improved thread synchronization and locality awareness. Its unique architecture achieves transfer bandwidths of up to 512 Gb/s for internal register transfers and 51.2 Gb/s for off-chip transfers. It implements 64, 64-bit cores and several levels of memory on chip, specifically 16 address-interleaved 128 kB L2 cache, 32 kB scratch pad memory and four external DDR3 memory controllers. The actual chip was shipped in the optimization strategy employed, termed cell-centred addressing that sequentially maps neighbour 3D cells to 2D cores by a scalar transformation vector which are then searched for migrating atoms in O (1) time. Such mappings are readily solved by classical algorithms but what makes the Godson-T and its progenies unique is the presence of on-chip hardware that implements a locality-aware parallel algorithm for data reuse and latency and congestion avoidance. Such issues typically arise from core to core communications when shared data structures are accessed in L2 cache or off-chip memory during the two-body force calculations. In this contribution, we present a short review of the current state of MD simulations with the aim of achieving speedups on a standalone computer. The discussion is expected to permit a reasonably easy starting point for standalone computational MD for the reader, with the view to readily scale such computers to larger clusters by virtue of interconnection. The work is intended to address single machines that support thread and message passing by virtue of the processor and memory architecture, by using any selection of proliferated tools which are freely available for such purposes. We discuss the rationales for the chosen approach of our own ongoing work in the hope that the generally high initial inertia could be overcome by the interested reader. Finally, we present some of our results thus far and point to an online repository where the evolving code is being maintained. We begin by outlining some standalone architectural configurations that are decidedly parallel or can be made so with relative ease.

1.2 Standalone architectures

The processor memory architectures of the traditional computer fall into one of two broad categories. In the first category, instruction fetch-execution cycles are repeated from the reset vector up until the last instruction is executed. The instructions and data are stored contiguously in the same, indistinguishable memory blocks. The instruction and data fetches happen along the same paths. In the second category, the instructions are generally stored in different memory locations and the paths to them are different. This sort of architecture is generally faster since simultaneous fetches and pre-fetches are possible. Several enhancements have been developed to support this behaviour, such as pipelining or super pipelining, which involve look-ahead methods and ways to avoid the problems that arise when branches are encountered in the linearly pre-fetched sequence of instructions and data. Such methods include branch prediction, cache memory for the frequently used data thereby saving on fetch times, and so on. A complete discussion of these techniques is beyond the present scope.

1.2.1 Classifications of parallelization paradigms

Parallel problems fall into two broad categories, embarrassingly parallel problems and serial problems. The placement of a problem depends on the extent and readiness to which it can be parallelized. A problem is said to be "embarrassingly parallel" if it is readily separated into identifiable, unique tasks that may then be executed separately. For such problems, the execution paths of the different threads are independent in the sense of the results of one thread not depending on any other thread. On the other hand, a "serial" problem is one that cannot be split into independent sub-problems and the results of one computation drive the next. Therefore, such a problem cannot easily be distributed across independent processing units without requiring interprocess communication. However, some problems can contain both elements and must be coded as combinations of embarrassingly parallel and serial. Most parallelization of interest occurs at the level of the iterations where the force fields are applied. Only these loops are considered for performance benchmarking. The other parts of the typical program are the user and output interfaces of a program. These parts of the code are concerned with loading particle structures into memory and can implement iterations that allow the particles in the configuration to relax to an equilibrium decided by the (new) boundary condition (called "setup" time), such as applied temperature. These parts of the code can take inordinately large multiples of the simulation time step, but typically occur only once per simulation and are therefore not considered in performance benchmarking [29].

1.3 Getting started with standalone computation

1.3.1 To code or not to code

For the innumerable material science problems that could be solved by MD techniques, there may or may not already exist a software, either specific for the class of the problem on hand or with an innate ability to be tuned for the problem. Many available packages are heuristic or "black box" in nature and cannot be dissected readily; others target a specific class of problems (through the available potentials and force fields), which introduce inflexibility towards new problems. Also, as mentioned elsewhere in this article, advancements in computer architecture and improved parallelization efficient and/or more accurate algorithms for the same software make it advantageous to develop software with high modularity in mind to keep up with developments in the field. Naturally, one then asks whether embarking on writing one's own code is a reasonable venture. To address this question meaningfully, careful considerations of the problem to be solved must be made, such as availability or access to similar software, the complexity of the problem if no software precedent exists, the technical resources available in terms of development time, target hardware and know-how, to highlight a few. An illustration of this scenario relates to density functional theory (DFT) calculations [40], for which numerous free software exist. However, different DFT software have different strengths and weaknesses. Many of them demonstrate applicability to the newer class of molecular and atomic aggregations found in the emerging field of nanotechnology, but perform poorly in many others. Next, the question would be whether the software is user adaptable to a new problem, such as ability to resolve small atom arrays of a totally different particle configuration and boundary conditions. For an academic undertaking, writing own software has obvious advantages and can lead to a software that is small, fast and potentially extendable to new classes of problems, particularly if the software is designed with modularity for code reuse in mind. Such codes can be understood and readily evolved by more developers with respect to parallelization efficiency and accuracy.

1.3.2 General tools

Standalone techniques that are considered to be successful are those that are highly scalable to implement larger systems while assuring/improving the desired levels of accuracy. Here, we define scalability as the ease and ability to increment the system size and computational power by physically adding similar systems to an existing system while adjusting only a small set of reconfiguration parameters. The closest parallel to this is the addition of a RAM module to upgrade system memory. Today, many interconnected systems are said to be "hybrid" since they are comprised of interconnections of subsystems that are heterogeneous with respect to hardware and operating system, sometimes referred to as a "farm". Interconnectivity is an obvious, fundamental tenet of the internet, where hardware compliance with controlled communication protocols and standards assures conformity, rather than the actual hardware and operating system in use. For this reason, the computing farm could readily support Apple or x86-based hardware, running Linux, BSD or Windows operating systems, all of them cooperating to reinforce the overall computational power. If one takes cognisance of the lower starting inertia and the need for simplicity for a better understood system, then the scaled system should ideally be homogeneous to inherently possess a plug-and-play character.

1.3.3 Parallelizable tools

Naturally, the question that can be asked at the onset is, how does one begin to exploit the full potential of a standalone computer having P processors and C cores? Second, is this set {P, C} available by default by virtue of the unit being on? Third, how does one benchmark the performance of this set? Interestingly, as suggested above, these questions are still largely unanswered. There are some inroads to the answers, and several studies have been conducted to attempt to correlate the performance of the system as a function of the size in terms of the set {P, C}. Peng et al. [25] measure the speedup on p cores for a problem of fixed size as $S(p) = T_1/T_p$, which is the ratio of the execution time on one core compared to the execution time on p cores. The strong-scaling parallel efficiency E_p is then $E(p) = S(p)/p$. They also monitor the performance metrics within cores for their quad-core simulations using Intel's VTune Performance Analyzer. Pal et al. [41] discuss the bottlenecks in parallelized MD computations and a hybrid algorithm using MPI with OpenMP threads for problems associated with the embedded atom model and Morse potentials. They contrast their study with similar studies using LAMMPS and find that their method produces enhanced performance. In attempting to contribute further clarity on these questions to a degree of usefulness and practicality, we begin by outlining some possible tools for the task. Many parallelizable tools are inherent or implicit in the programming environment and their specific support libraries, allowing the user code to switch to parallel computation with relative ease. There are also tools that allow benchmarking of the performance of the system relative to a standard. One such standard known as LINPACK [42, 43], for instance, times the co-factorization of a real, dense matrix. Other computational tasks that have been used as tests are solvers of ordinary differential equations (ODEs), fast Fourier transform (FFT), sparse matrix algorithms and others. Therefore, the true performance of a system needs to be considered holistically as an aggregate of several benchmark tests to obtain a more rounded feel for the relative computational ability of the system. The examples given here are based on the C-programming language which can be run on many common compilers with standardized libraries in certain environments and platforms, specifically Linux and Windows. We also briefly consider packages such as MATLAB that are now gradually moving towards parallel computing in line with the global trend. In the past, the distributed memory model on most computers, including some parallel computers, was incompatible with the "Matrix Laboratory" or MATLAB memory model. As a result, many operations from within MATLAB could not readily be parallelized. At the same time, the demand for such features in MATLAB was low [44]. This situation has changed significantly and MATLAB has evolved into an environment that actively supports large-scale projects for multi-core, multiprocessor architectures, which take it far beyond its traditional role of handling matrix operations. By extension, MATLAB has evolved to a form referred to as parallel MATLAB that can handle networked clusters by implementing three kinds of parallelism. First, there is multithreaded parallelism (MP), where an instance of MATLAB generates several simultaneous instruction streams. On a multi-core or multiprocessor machine, the streams are distributed to the set $\{P, C\}$ and executed. An example of an operation that may benefit from MP is summation of the elements of a matrix. Second, there is distributed computing (DC), where several instances of MATLAB run multiple independent operations on separate computers with separate memories. Such an application of MATLAB is considered trivial to implement and yet offers clear advantages of parallel execution. The DC involves a single code being run many times with different parameters or random number seeds. Third, there is explicit parallelism (EP), which requires newer programming approaches that consist of parallel loops, distributed arrays and the program code runs on several processors or computers that can have separate memories. In summary, choosing the particular implementation of parallelism can be a difficult task, and one needs to weigh the task to be solved against the hardware in place.

1.4 Parallel processing paradigms in the C-language

1.4.1 Threads and message passing

The C-language is native to Unix-based systems, but its high portability has made it standardized and widely available across several platforms. Although there are newer contenders to parallelization, but the C-language is still widely used for parallelization for three main reasons. First, complex data structures that closely resemble collections of structured particles in a system can be defined, referenced and manipulated within memory with considerable ease. The basic data structure typically consists of a particle type object with directly accessible attributes such as mass, position and velocity. Entire collections of such particles, which represent particle configurations within a domain, can be manipulated just as readily as its individual particles through structure access functions. Second, the use of pointers within the language introduces a flexibility and enabling modularity. In this way, individual particles or entire collections of particles can be passed on between different sections the code with ease. An instance where this flexibility is called for is during particle migration. This requires that a particle is added or deleted from a collection. Third, the language is fast and has a number of excellent heap memory management functions. It is native and highly popular on Unix-based systems. The main idea behind thread processing as exploited in parallel programming involves a single problem process giving rise or "spawning" multiple, dynamically managed sub-programs called "threads." The threads run concurrently and independently and are structured such that they can access shared memory. Shared memory access can involve some interthread communication, requiring careful management of all spawned threads, from the moment each is spawned to the moment it terminates. The process of killing a thread releases its resources for reassignment by the manager. A critical role of the manager is to implement careful synchronization of threads to avoid thread collisions and racing conditions by assuring a mutual exclusivity of memory access during a focus by a particular thread. The concept of threads is well-established in the C-programming language and is conveyed by a number of paradigms. For instance, the POSIX threads standard, or p-threads, makes use of a standardized library header "pthread.h," which contains the necessary definitions of constants, functions and types. The code in Figure 1.1 spawns three threads using this approach. Other popular parallelization techniques available to C-programs include the multiprocessing application programming interfaces (APIs), of which the open multiprocessing API (OpenMP) is a popular implementation, and various message passing interfaces (MPIs). The MPIs are libraries that define code and routines which run on a diverse cross-section of platforms including Linux and Windows. The most commonly referenced and freely available MPIs are OpenMPI (not to be confused with OpenMP), MPI, MPICH2 and LAM-MPI. Each of these MPIs may be considered a cross between p-threads and OpenMP.

1.4.2 Open multiprocessing programming

A significant amount of time in complex, repetitive calculations is spent executing loop statements. These iterations lend themselves readily to parallelization using OpenMP by the inclusion of the compiler directive #pragma followed by the code to be parallelized. An example is shown in Figure 1.2. The same directive can be used to

```
#include <stdio.h>
#include <stdlib.h>
#include <pthread.h>
#define NTHREADS 5
void *myFun(void *x){
  int tid;
  tid = *((int *) x);
  printf("Executed thread: %d!\n", tid);
  return NULL;
}
  int main(int argc, char *argv[]){
  pthread_t threads[NTHREADS];
  int thread_args[NTHREADS];
  int rc, i;
  for (i=0; i<NTHREADS; ++i)
    {
      thread_args[i] = i;
      printf("spawning thread %d\n", i);
      rc = pthread_create(&threads[i], NULL, myFun, (void *)&thread_args[i]);
    }
  /* wait for threads to finish */
  for (i=0; i<NTHREADS; ++i) {
    rc = pthread_join(threads[i], NULL);
  }

  return 1;
}
```

Figure 1.1: An example of p-thread code spawning three threads.

```
#include <stdio.h>
#include <omp.h>

int main(int argc, char **argv)
{
  int i, thread_id, nloops;

#pragma omp parallel private(thread_id, nloops){
    nloops = 0;
#pragma omp for
    for (i=0; i<1000; ++i){
        ++nloops; }
    thread_id = omp_get_thread_num();
    printf("Thread %d, loop iteration %d.\n", thread_id, nloops );
  }
  return 0;
}
```

Figure 1.2: An example of the OpenMP implementation for parallelization.

specify the aspects of critical code, i. e. code that is common to all the threads, as shown in Figure 1.3. The critical code can, for instance, manage a variable that is global relative to the other parallelized code sections. This approach is implicitly serial and has the virtue of eliminating the problems caused by collisions where code sections try to access the same variable simultaneously.

```
#include <stdio.h>
#include <omp.h>

int main(int argc, char *argv[])
{
    int i, thread_id;
    int glob_nloops, priv_nloops;
    glob_nloops = 0;
    #pragma omp parallel private(priv_nloops, thread_id)
    {
        priv_nloops = 0;
        thread_id = omp_get_thread_num();

        #pragma omp for
        for (i=0; i<100000; ++i){
            ++priv_nloops;}
        #pragma omp critical{
            printf("Thread %d is adding its iterations (%d) to sum (%d), ",
                   thread_id, priv_nloops, glob_nloops);
            glob_nloops += priv_nloops;
            printf(" total nloops is now %d.\n", glob_nloops);}
    }
    printf("Total # loop iterations is %d\n",
           glob_nloops);
    return 0;
}
```

Figure 1.3: An example of ordinary OpenMP and critical OpenMP processes.

```
#pragma omp parallel private(priv_nloops, thread_id) _
                     reduction(+:glob_nloops)
```

Figure 1.4: Syntax of the compiler directive that invokes output reduction of parallel processes.

The process of combining the outputs of several smaller parallel sections is known as reduction. In OpenMP, there is a separate keyword that is used with the #pragma omp directive to carry out more reduction efficiently because of its importance. The syntax is shown in Figure 1.4.

There is a multitude of other keywords that are associated with the #pragma omp directive, for the streamlining of various multiprocessing tasks.

Figure 1.5 shows the specific application of the authors to which parallelization was applied. The code snippet is only a small part of a bigger program and shows the calculation of the total potential energy using the Sutton–Chen implementation of the Finnis–Sinclair potential.

1.4.3 Message passing interface programming

The MPI has several implementations that support cross-platform multiprocessing. Through this interface, the program assigns a collection of computational nodes

```
// mpicc go_mpi.c -o go_mpi
// mpirun -n 4 go_mpi

#include <stdio.h>
#include <mpi.h>

int main(int argc, char *argv[]){
  int myrank, nprocs;

  MPI_Init(&argc, &argv);
  MPI_Comm_size(MPI_COMM_WORLD, &nprocs);
  MPI_Comm_rank(MPI_COMM_WORLD, &myrank);

  printf("This is node %d of %d\n", myrank, nprocs);

  MPI_Finalize();
  return 0;
}
```

Figure 1.5: An example of a program parallelized using MPI.

resources such as memory and coordinates communication and synchronization with a set of processes running on the nodes with ease. In this approach, any computing node is handled identically. Thus, the interface does not distinguish the elements in the set {P, C}, but ropes each element as a computational resource. This approach is attractive because it operates well across clusters. The scalability is then a matter of increasing the number of such interconnects. Various supercomputers on the planet today are implemented by such arrangements, with some using proprietary interconnects and others with custom, well-guarded interconnects. A typical example of MPI code running on a Linux cluster is shown in Figure 1.5.

1.4.4 The GPU approach

GPUs are increasingly being used to implement compute nodes due to their vast processing power as a consequence of sheer speed and their reduced instruction set computing (RISC) architectures and low power consumption [45–47]. GPUs are specialized at certain operations requiring massive amounts of data at blazing speeds, at relatively low cost. Many GPUs carry multiple cores and are naturally suited to the kinds of calculations that are common in advanced physics and engineering. Multi-GPUs each on a native board known as blades daughter boards can be clustered together on bigger motherboards. Thus, several blades can implement high-performance compute clusters while taking up very little space. Understandably, the power electrical power demands of such clusters need to be carefully considered, and such clusters may be housed specially in localities referred to as render farms. Brown et al. [48, 49] discuss some issues of porting a large MD for CPUs onto parallel hybrid machines based on GPUs. They also use hybrid decomposition on code intended for distributed memory and accelerator cores with shared memory and efficient code porting of short-range force calculations through an accelerated programming model within an existing LAMMPS MD program with a custom library called Geryon. Their technique is based on load balancing of work between the CPU and accelerator cores and can be benefit cases that require the processing power of CPU cores over GPUs. This is still necessary because many GPU algorithms currently cannot achieve peak floating-point performance at present as measured relative to wall-clock time when compared to CPUs. This is a consequence of the specialization of the GPU's specialization of certain operations and not others. Having some operations run on CPUs can minimize the amount of coding required for functions executed on GPUs in a hybrid system [48–52]. For instance, Brown et al. [48] show that double-precision performance can be poorer than single or mixed precision in such clusters, indicating that further optimization is required. Other workers [10] argue that increased modularity is advantageous for evolvable scientific software. The goal of writing the software is to assure functionality alongside while carefully considering parallelization efficiency and most importantly, accuracy and reproducibility of results relative to empirical results that may exist.

1.4.5 Cloud virtualization

In very recent years, the concept of the computing Cloud has arose initially as a means of getting large, safe and storage facilities for organization files. This has gradually been progressed to include outright distributed computation and large-data crunching. For instance, financial markets generate vast amounts of time-series data during a trade. The needs of forecasting, fraud detection, internet trade, etc. all require an almost real-time computational element. Depending on the intended functionality and application of the remote user, these powerful new features may be supplied in a limited way free of charge, or at a nominal fee. In any case, this current trend absolves the user from making any major financial outlay in terms of the starting hardware while reaping the benefits of powerful compute facilities and storage. However, security and data privacy and a full grasp of the trappings of this relatively young concept remains a concern for large corporations with obvious interest to safeguard their operations. One recurrent feature of such remote compute facilities is the ability to define virtual environments familiar to the user through virtualization. In essence, several instances of a virtual high-performance computer can be created and loaded by a simple script or graphical interface from a simple less powerful one serving as a terminal from across the planet. The instances can be either homogenous or heterogeneous in terms of the loaded operating system. Examples of organizations with such offerings are Yellow Circle [53], Microsoft Cloud (Azure) [54],

Google Cloud platform [55] and others that support free virtualization for individuals in a research university through a secure account. However, the free accounts suffer the limitation in the maximum number of possible virtual instances, a drawback not suffered by paying users. In addition, the technical expertise required to have tasks up and running on such systems with carefully enumerated set {P, C} is essentially nil since server side support exists to tailor applications, particularly for the paying users. In the case of Yellow Circle, for instance, the user can remotely configure the network topology by designing routers, internet protocol addresses, subnets and so on, all through a simple graphical user interface. An image capture of the configuration screen on Yellow Circle with a free account is shown in Figure 1.6. The tasks of assigning compute nodes, assimilating final output data gathering and task scheduling and general network management are equally straightforward. Consequently, clouds may indeed be the ultimate future of cluster computing. Cloud virtualization can improve the utilization of compute resources while reducing implementations costs. It also introduces a desirable ability to a user to customize their system without many of the technical challenges experienced in traditional grids and cluster configuration. However, at present, virtualization-based cloud has limited performance in its infancy and is yet limited for HPC for a number of reasons. One such limitation can be caused by system virtual machine monitors for supervisory control for the cores and virtual memory peripherals. Methods therefore have to be found to minimize such delays. Ren et al. [56] propose a unique, lightweight supervisory cloud infrastructure called nOSV that serves both HPC and normal applications concurrently. The environment constructs high-performance virtual machines with dedicated resources which operate without supervisory interference, thereby keeping the performance of the

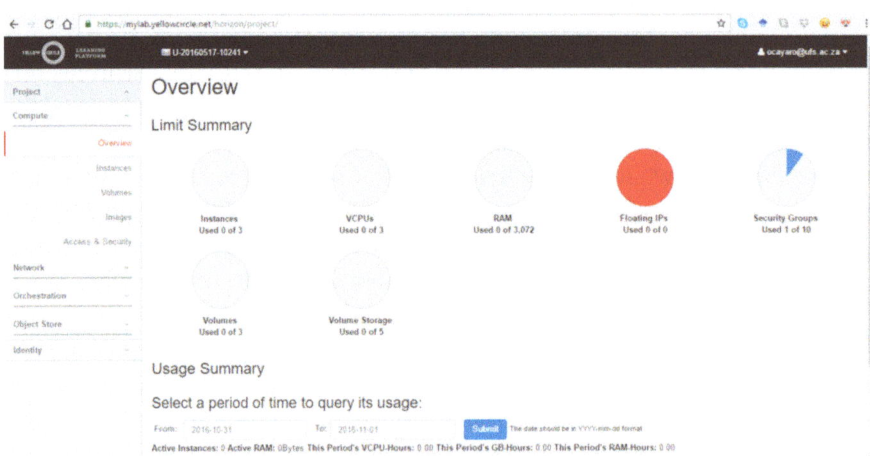

Figure 1.6: A remote virtualization cloud implemented in Yellow Circle for computational purposes.

traditional Cloud. With their implementations, they report significant performance improvements over other similar virtualization environments.

1.5 Summary of results

Our own work pertains to face-centred cubic (fcc) atomic structures of metallic atoms, such as copper atoms. The force fields were derived from the embedded atom model adaptation by Sutton–Chen of the Finnis–Sinclair potential. To illustrate the speedup, we calculated the equilibrium lattice parameter (a), the vacancy formation energy of copper, where an atom is removed for just below the surface of the bulk structure and taken to a point several hundred atomic radii away from the surface (to simulate a point at "infinity"), with a cut-off distance of 10 lattice parameters. Similarly, we calculate the extraction energy [7, 9]. The bulk array consists of 30,000 atoms, and computation time is measured using system time functions that are outside the particle setup, the equilibration steps and the final data output steps. The same timer functions are used for the non-parallelized iterations as well as the parallelized code. The configured hardware is the Dell Optiplex 3010 mentioned above. Without parallelization, the run times in a calculation of E_v on an Ubuntu Linux terminal were 6.70 hours and 5.21 hours for the deterministic and Monte-Carlo simulations, respectively. The maximum parallelized speedups are shown in Table 1.1. The results show a speedup of 1.8 on the wall-clock time for 5 % accuracy on the lattice parameter. The Monte-Carlo method exhibited higher speedup, with the lower acceptance rate, much higher than observed for the deterministic, MD-based simulation. A version of the software can be obtained from [57].

Table 1.1: Parameter calculations and on 30,000 atom fcc copper array using thread-based parallelization.

Method	a(nm)	E_v(eV)	E_{ext}(eV)	Speedup
Deterministic	0.363	1.31	4.39	1.8
Monte-Carlo (30 % acceptance)	0.374	1.66	5.84	2.2

1.6 Conclusions

This article discusses the considerations to embark on computational research in MD using optimized multi-core computers that can also be operated as a standalone computer, or a node in cluster-based computing. At the onset, a fundamental goal was in the establishment of a computing facility that exploits the features of the single machine, with the secondary aim to extend the functionality to a cluster of such machines, making low-cost HPC possible for the group. The article presents a survey of the current state of the art, on which we base our own ongoing work. We consider what it takes to achieve parallelism and efficiency, providing arguments and pointers to aspects to consider if the need to write own as opposed to using third-party software should arise. Furthermore, we discuss the growing trends in high-performance computation, from the use of GPUs to internet-based virtualization on computing clouds. Finally, we present the summarized results for a sizable array of copper atoms. Both MD and Monte-Carlo methods showed speedup when the iterative steps were parallelized using threads on four cores. A link to a version of the evolving code in a public repository, called the Verlet–Stormer velocity program code [57], is also given for the interested reader. Since threads and MPI are concepts that are standardized and supported in libraries across many platforms, the C-program on which the present work is based could readily be ported to other systems.

References

[1] Horsch M, Vrabec J, Bernreuther M, Grottel S, Reina G, Wix A, et al. Homogeneous nucleation in supersaturated vapors of methane, ethane, and carbon dioxide predicted by brute force molecular dynamics. J Chem Phys. 2008;128:164510.

[2] Mendelev MI, Han S, Srolovitz DJ, Ackland GJ, Sun DY, Asta M. Development of new interatomic potentials appropriate for crystalline and liquid iron. Philosophical Magazine. 2003;83 (35):3977–3994. DOI:10.1080/14786430310001613264.

[3] Sutton AP, Chen J. Long-range Finnis–Sinclair potentials. Philos Mag Lett. 1990;61(3):139–156.

[4] Das A, Ghosh MM. MD simulation-based study on the melting and thermal expansion behaviours of nanoparticles under heat load. Computational Mater Sci. 2015;101:88–95. DOI:10.1016/j.commatsci.2015.01.008.

[5] Van Der Walt C, Terblans JJ, Swart HC. Molecular dynamics study of the temperature dependence and surface orientation dependence of the calculated vacancy formation energies of Al, Ni, Cu, Pd, Ag, and Pt. Computational Mater Sci. 2014;83:70–77. DOI:10.1016/j.commatsci.2013.10.039.

[6] Abraham MJ, Murtolad T, Roland Schulz R, Palla S, Jc J, Hessa B, et al. GROMACS: high performance molecular simulations through multi-level parallelism from laptops to supercomputers. SoftwareX 2015. DOI:10.1016/j.softx.2015.06.001.

[7] Ocaya RO, Terblans JJ. to appear in Journal of Physics: Conference Series. Paper presented at CCP2016-28th IUPAP Conference on Computational Physics; Johannesburg, South Africa 2016 July 10–14. 2016 July 10–14 Temperature specification in atomistic molecular dynamics and its impact on simulation efficacy 2017.

[8] Ocaya RO, Terblans JJ. C-language package for standalone embedded atom method molecular dynamics simulations of fcc structures. SoftwareX. 2016;5:107–111. DOI:10.1016/j.softx.2016.05.005.

[9] Ocaya RO, Terblans JJ. to appear in Journal of Physics: Conference Series. Paper presented at CCP2016-28th IUPAP Conference on Computational Physics; Johannesburg, South Africa 2016 July 10–14. 2016 July 10–14 Coding considerations for standalone molecular dynamics simulations of atomistic structures 2017.

[10] Buchholz M, H-J B, Vrabec J. Software design for a highly parallel molecular dynamics simulation framework in chemical engineering. J Comput Sci. 2011;2:124–129. DOI:10.1016/j.jocs.2011.01.009.

[11] Smirnov BM. Energetics of clusters with a face centered-cubic structure. Zh Eksp Teor Fiz. 1995;107:2080–2091.

[12] Terblans JJ. Calculating the bulk vacancy formation energy (Ev) for a Schottky defect in a perfect Cu(111), Cu(100) and a Cu(110) single crystal. Surf Interface Anal. 2002;33:767–770. DOI:10.1002/sia.1451.

[13] Mattsson TR, Mattsson AE. Calculating the vacancy formation energy in metals: Pt, Pd, and Mo. Phys Rev B. 2002;66:214110.

[14] Sebastian IS, Aldazabal J, Capdevila C, Garcia-Mateo C. Diffusion simulation of CrFe bcc systems at atomic level using a random walk algorithm. Phys Stat Sol (A). 2008;205(6):1337–1342. DOI:10.1002/pssa.200778124.

[15] Jian-Min Z, Fei M, Ke-Wei X. Calculation of the surface energy of fcc metals with modified embedded-atom method. Appl Surf Sci. 2004;13(7):34–42. DOI:10.1016/j.apsusc.2003.09.050.

[16] Griebel M, Knapek S, Zumbusch G. Numerical simulation in molecular dynamics. Texts in computational science and engineering 5. Berlin: Springer, 2007978-3-540-68094-9.

[17] Car R, Parrinello M. Unified approach for molecular dynamics and density functional theory. Phys Rev Lett. 1985;55(22):2471–2474.

[18] Car R, Parrinello M, Loubeyre PP, Boccara N. The unified approach for molecular dynamics and density functional theory. In: Loubeyre P.P., Boccara N., editors. Simple molecular systems at very high density. Vol. 186 of NATO ASI Series, Series B, Physics. New York: Plenum Press, 1989:455–476.

[19] Remler DK, Madden PA. Molecular dynamics without effective potentials via the Car–Parrinello approach. Mol Phys. 1990;70(6):921–966.

[20] Tuckerman ME. Ab initio molecular dynamics: basic concepts, current trends and novel applications. J Phys: Condens Matter. 2002;14 R1297R1355 Online at stacks.iop.org/JPhysCM/14/R1297.

[21] Moore GE. Cramming more components onto integrated circuits. Electronics Magazine 1965; 38(8):114–117.

[22] Moore GE. Cramming more components onto integrated circuits. Proc IEEE. 1998;86(1):82–85.

[23] Dongarra J. Paper presented in International Supercomputer Conference. Germany: Heidelberg, 2004 June Survey of present and future supercomputer architectures and their interconnects 2225.

[24] Intel Corporation. document on the Internet;, January 2016. cited Available from. 2017 January 14 http://ark.intel.com/products/65703.

[25] Peng L, Tan G, Kalia RK, Nakano A, Vashishta P, Fan D, et al. Preliminary investigation of accelerating molecular dynamics simulation on Godson-T many-core processor. Euro-Par 2010 Parallel Processing Workshops. 2010;6586:349–356.

[26] Peng L, Tan G, Kalia RK, Nakano A, Vashishta P, Fan D, et al. Preliminary investigation of accelerating molecular dynamics simulation on godson-T many-core processor. Parallel Processing Workshops, Euro-Par 2010: HeteroPar 2010, HPPC 2010, HiBB 2010, CoreGrid 2010, UCHPC 2010, HPCF 2010, PROPER 2010, CCPI 2010, VHPC 2010, Ischia; Italy; 2010. 2010 August 31–September 3 Code 86021.

[27] Peng L, Tan G, Kalia RK, Nakano A, Vashishta P, Fan D, et al. Preliminary investigation of accelerating molecular dynamics simulation on Godson-T many-core processor. Vol. 6586 of the Series Lecture Notes in Computer Science Euro-Par. 2010;2010:349–356.

[28] Bernreuther M, Bungartz H-J, Hülsemann F, Kowarschik M, Rüde U. Proceedings of the 18th Symposium Simulations Technique (ASIM2005), in Frontiers in Simulation, SCS European Publishing House; 2005;117–123 15.

[29] Mangiardi CM, Meyer R. A hybrid algorithm for parallel molecular dynamics simulations [document on the Internet]. 2016 [cited 2016 October 31]. Available from: https://arxiv.org/abs/1611.00075.

[30] Mangiardi CM, Meyer R Molecular-dynamics simulations using spatial decomposition and task-based parallelism. In: Bélair J., Frigaard I., Kunze H., Makarov R., Melnik R., Spiteri R., editors. Mathematical and computational approaches in advancing modern science and engineering. Switzerland: Springer International, 2016:133–140.

[31] Meyer R. Efficient parallelization of short-range molecular dynamics simulations on many-core systems. Phys Rev E. 2013;88(5):053309.
[32] Plimpton S. Fast parallel algorithms for short-range molecular dynamics. J Comput Phys. 1995;117(1):1–19.
[33] Todorov IT, Smith W, Trachenko K, Dove MT. DL POLY 3: new dimensions in molecular dynamics simulations via massive parallelism. J Mater Chem. 2006;16(20):1911–1918.
[34] Phillips JC, Braun R, Wang W, Gumbart J, Tajkhorshid E, Villa E, et al. Scalable molecular dynamics with NAMD. J Comput Chem. 2005;26(16):1781–1802.
[35] Ackland GJ, D'Mellow K, Daraszewicz SL, Hepburn DJ, Uhrin M, Stratford K. The MOLDY short-range molecular dynamics package. Comput Phys Commun. 2011;182(12):2587–2604.
[36] Berendsen HJC, Van Der Spoel D, Van Drunen R. GROMACS: a message-passing parallel molecular dynamics implementation. Comput Phys Commun. 1995;91(1):43–56.
[37] Needham PJ, Bhuiyan A, Walker RC. Extension of the AMBER molecular dynamics software to Intel's Many Integrated Core (MIC) architecture. Comput Phys Commun. 2016;201:95–105.
[38] Hager G, Wellein G. Introduction to high performance computing for scientists and engineers. Boca Raton, FL: Chapman and Hall, 2011.
[39] Wang J, Gao X, Li G, Q L, Hu W, Chen Y. Godson-3: a scalable multicore RISC processor with x86 emulation. IEEE Micro. 2009;2(29):17–29. DOI:10.1109/MM.2009.30.
[40] Jain A, Shin Y, Persson KA. Computational predictions of energy materials using density functional theory. Nature Reviews Materials. 2016;1. Article number: 15004. DOI:10.1038/natrevmats.2015.4.
[41] Pal A, Agarwala A, Raha S, Bhattacharya B. Performance metrics in a hybrid MPI–OpenMP based molecular dynamics simulation with short-range interactions. J Parallel Distrib Comput. 2014;74:2203–2214.
[42] Petitet A, Whaley RC, Dongarra J, Cleary A. HPL - A portable implementation of the high-performance linpack benchmark for distributed-memory computers [document on the internet]. 2016;[cited Available from. 2016 October 22 http://www.netlib.org/benchmark/hpl/.
[43] The Linpack Benchmark [document on the Internet]. October 2016. [cited Available from. 2016 October 22 https://www.top500.org/project/linpack/.
[44] Cleve M. Parallel MATLAB: multiple processors and multiple cores [document on the Internet]. October 2016. [cited Available from 2007 January 01] https://www.mathworks.com/company/newsletters/articles/parallel-matlab-multiple-processors-and-multiple-cores.html.
[45] Govender N, Wilke DN, Kok S, Els R. Development of a convex polyhedral discrete element simulation framework for NVIDIA Kepler based GPUs. J Comput Appl Math. 2014;270:386–400. DOI:10.1016/j.cam.2013.12.032.
[46] Govender N, Wilke DN, Kok S, Els R. Collision detection of convex polyhedra on the NVIDIA GPU architecture for the discrete element method. Appl Math Comput. 2015;267:810–829. DOI:10.1016/j.amc.2014.10.013.
[47] Govender N, Rajamani RK, Kok S, Wilke DN. Discrete element simulation of mill charge in 3D using the BLAZE-DEM GPU framework. Minerals Eng. 2015;79:152–168. DOI:10.1016/j.mineng.2015.05.010.
[48] Brown WM, Wang P, Plimpton SJ, Tharrington AN. Implementing molecular dynamics on hybrid high performance computers – short range forces. Comput Phys Commun. 2011; 182(4):898–911. DOI:10.1016/j.cpc.2010.12.021.
[49] Brown WM, Kohlmeyer A, Plimpton SJ, Tharrington AN. Implementing molecular dynamics on hybrid high performance computers – Particle-particle particle-mesh. Comput Phys Commun. 2012;183(3):449–459. DOI:10.1016/j.cpc.2011.10.012.
[50] Stone JE, Phillips JC, Freddolino PL, Hardy DJ, Trabuco LG, Schulten K. J Comput Chem. 2007;28:2618–2640.

[51] Schmid N, Botschi M, Van GWF. J Comput Chem. 2010;31:1636–1643.
[52] Hampton S, Alam SR, Crozier PS, Agarwal PK. Proceedings of the 2010 International Conference on High Performance Computing and Simulation (HPCS 2010), 2010.
[53] Yellow Circle [document on the Internet. October 2016. [cited Available from. 2016 October 29 https://mylab.yellowcircle.net.
[54] Microsoft Azure [document on the Internet]. 2016;[cited Available from. 2016 October 29 https://azure.microsoft.com/en-us/.
[55] Google Cloud Platform [document on the Internet]. October 2016. [cited Available from: 2016 October 29 https://cloud.google.com/.
[56] Ren J, Qi Y, Dai Y, Xuan Y, Shi Y. nOSV: a lightweight nested-virtualization VMM for hosting high performance computing on cloud. J Syst Softw. 2017;124:137–152. DOI:10.1016/j.jss.2016.11.001.
[57] VSV Software [software on the Internet]. 2016;[cited Available from: 2016 October 29 https://github.com/ElsevierSoftwareX/SOFTX-D-15-00054.

Suman Chowdhury, Arka Bandyopadhyay, Namrata Dhar
and Debnarayan Jana

2 Optical and magnetic properties of free-standing silicene, germanene and T-graphene system

Abstract: The physics of two-dimensional (2D) materials is always intriguing in their own right. For all of these elemental 2D materials, a generic characteristic feature is that all the atoms of the materials are exposed on the surface, and thus tuning the structure and physical properties by surface treatments becomes very easy and straightforward. The discovery of graphene have fostered intensive research interest in the field of graphene like 2D materials such as silicene and germanene (hexagonal network of silicon and germanium, respectively). In contrast to the planar graphene lattice, the silicene and germanene honeycomb lattice is slightly buckled and composed of two vertically displaced sublattices.The magnetic properties were studied by introducing mono- and di-vacancy (DV), as well as by doping phosphorus and aluminium into the pristine silicene. It is observed that there is no magnetism in the mono-vacancy system, while there is large significant magnetic moment present for the DV system. The optical anisotropy of four differently shaped silicene nanodisks has revealed that diamond-shaped (DS) silicene nanodisk possesses highest static dielectric constant having no zero-energy states. The study of optical properties in silicene nanosheet network doped by aluminium (Al), phosphorus (P) and aluminium-phosphorus (Al-P) atoms has revealed that unlike graphene, no new electron energy loss spectra (EELS) peak occurs irrespective of doping type for parallel polarization. Tetragonal graphene (T-graphene) having non-equivalent (two kinds) bonds and non-honeycomb structure shows Dirac-like fermions and high Fermi velocity. The higher stability, large dipole moment along with high-intensity Raman active modes are observed in N-doped T-graphene. All these theoretical results may shed light on device fabrication in nano-optoelectronic technology and material characterization techniques in T-graphene, doped silicene, and germanene.

Keywords: density functional theory, germanene, optical properties, silicene, T-graphene

2.1 Introduction

The recent developments of two-dimensional (2D) materials [1] have fostered a great deal of research interest since the first isolation of graphene [2–5]. The emergence of each new material brings excitement as well as puzzles in their characterization and

https://doi.org/10.1515/9783110467215-002

physical properties. These 2D materials offer an unusual platform for predicting various heterostructures suitable for versatile applications. The properties of these materials are usually distinctly different from those of their 3D counterparts. Besides, these characteristic 2D materials offer reasonable flexibility in terms of tailoring their electronic properties.

Graphene is also regarded as the building block of other various allotropes of carbon materials [6]. Till today, most of these materials exist in the hypothetical world, although several interesting theoretical works reveal various physical (electronic, magnetic, optical and transport) properties of these materials [7–11]. These theoretical works naturally facilitate further insights into experimental synthesis and investigations of these materials. It is natural to think about the existence of graphene-like hexagonal network of group IV elements due to similar electronic configuration. However, it is only silicene that has been synthesized successfully on Ag(111) substrate [12, 13, 15, 16], Ir(111) substrate [17] and also on conductive ceramic $ZrB_2(0001)$ [18]. Recent density functional theoretical (DFT) calculations [19] predict two different buckling of silicene, one is the low buckling (buckling height ~ 0.44 Å) and another one is high buckling (buckling height ~ 2.15 Å); however, the low-buckled (LB) structure is observed to be more stable than the high-buckled structure. Presence of buckling clearly indicates a tendency towards sp^3 hybridization from pure sp^2 hybridization associated with planar geometry. The most illuminating phenomenon in silicene is the emergence of linear electronic band structure like that of the graphene in contrast to the bulk phase of silicon [8, 20–22]. The linear band structure is the manifestation of long-range hexagonal periodicity and sublattice symmetry associated with the silicene. In contrast to graphene, bandgap can be tuned in silicene by applying transverse external electric field [23–25] without the need for any chemical functionalization in contrast to graphene.

Compared to graphene, in silicene, the spin–orbit interaction is quite prominent that eventually gives rise to small bandgap (~1.55 meV) opening at Dirac point [26, 27]. Notably, this bandgap value is far from present device application point of view, although it emerges a new field of research interest regarding the phenomena of quantum spin Hall effect (QSHE) [27, 28]. The QSHE is one of the recent scientific attraction due to its importance for technological applications in the fields of spintronics and 2D nanotechnology. Investigation about density of states (DOS) can help to explain metallic, semi-metallic, semiconductor or insulating behaviour of a system. Besides, in contrast to graphene, which shows QSHE, one can indeed observe QSHE in silicene at relatively higher temperature. Moreover, unlike graphene, silicene is a topological insulator characterized by a full insulating gap in bulk with a helical gapless edge. Coupled effect of transverse tunable external field with the spin–orbit interaction can transform silicene from topological insulator to band insulator [10]. Chemical functionalization to silicene enhances the possibility of its application in different branches. Hydrogenation as well as fluorination of silicene opens up a bandgap, although fluorinated silicene is observed to be more stable than the hydrogenated one [29]. Ciraci et al. [30] have extensively studied many aspects of silicene and stabilization properties with respect to defects. These theoretical results validate the promising usefulness of silicene in future device applications, and it is also interesting to note that silicene is one step ahead compared to graphene due to its compatibility with the present matured silicon-based semiconductor technology.

While experimental studies take time and require a lot of resources, computational study, on the other hand, particularly the first-principles calculations, are playing an important key role in engineering the bandgap, scanning tunnelling microscopy (STM) images and the stability of the proposed 2D structures. DFT has been employed in order to understand the structural and electronic properties of silicene on a series of metallic and semiconducting substrates. Unlike graphene, silicene cannot grow in free-standing (FS) form and for its growth, suitable single metal substrate is required [31]. FS silicene is known to be a LB structure with hybridization between sp^2 and sp^3 due to the large Si–Si interatomic distance that weakens the π–π overlap [26].

FS silicene being a LB structure, it may be possible to favor sp^3 hybridization through Si–H bonding with enhanced buckling. With this motivation, the fully hydrogenation of silicene $(2\sqrt{3}\times 2\sqrt{3}R30^o)$ has been explored [32, 33] by DFT calculation and STM images. As a result, synthesis of half silicane (one Si sublattice is fully H-saturated and the other sublattice remains intact, forming a perfect 1×1 structure) will trigger the investigation regarding other unexplored electronic and optical properties. The scanning tunnelling spectroscopy (STS) measurements suggest that the DOS of hydrogenated silicene is substantially decreased from the E_F to 0.6 eV compared to silicene, indicating an opening of gap. This pioneering work will be useful for controllable hydrogen storage. Note that, in this respect, theoretical results obtained for FS silicene will also be applicable in the presence of a weakly interacting substrate such as semiconducting Al_2O_3. A sizable bandgap without disturbing the electronic characteristics of silicene is particularly fruitful for field effect transistors (FETs) [34].

Like 2D materials: such as hexagonal boron-nitride (h-BN), silicene, transition metal dichalcogenides (TMDs) having general structure of MX_2 where M and X denote TM atom and oxygen family elements, respectively [35, 41]. In the cases of h-BN and H-passivated Si–SiC(0001), the effects of the interaction are close to negligible so that quasi-FS silicene [36] may be realized.

These novel materials possess extraordinary characteristics features such as high carrier mobility, high thermal conductivity, Dirac cone-like features in band structures [37], etc. Besides, it is possible to apply chemical functionalization technique effectively in these 2D sheets because of their high surface to volume ratio. Emergence of elemental graphene-like sheet of germanium (Ge), called germanene, after graphene and silicene, has enriched this 2D materials family by its unique electronic and optical properties [19, 31, 38].

The customary issue with graphene that disavow its potential functionality in electronic devices is its zero bandgap. This hurdle can be solved by reducing symmetry of the system which affects the Π–Π^* bonds. In a study [39], Enyashin et al. have investigated the structural and electronic properties of 12 differently hybridized, artificial 2D graphene allotropes. The underlying motivation was obviously the formidable ability of the C atom to form a variety of stable allotropes which exhibit various structure-dependent physical properties. Different degrees of hybridization for these 12 structures have been expressed as sp^n, classification of the linkage based on valence orbital hybridization (n) (i.e. for mixed state $1 < n < 3$, $n \neq 2$) [40]. One of the important structures known as tetragonal graphene, TG in short, although having thermodynamically stable metastable state, has attracted the researchers for further studies.

Although application of 2D materials like silicene, germanene, TG sheet, etc. beyond graphene is still very limited, but their outstanding properties can afford new excellent opportunities to the researchers in this area [5, 10, 41]. In Table 2.1, we have provided some important structural parameters of silicene, germanene and three conformers of TG sheet (planar, buckled and C_4).

Table 2.1: Some important parameters of silicene, germanene and planar (P), buckled (B), C_4 T-graphene sheet.

Parameters	Silicene	Germanene	T-graphene
Lattice constant a(Å)	3.858	4.06	3.42 [159] / 3.47 (P) [39], 4.84 (B) [159], 3.447 (C_4) [167]
Bond length d(Å)	2.232	2.341	1.417 [159] / 1.429 (P) [39], 1.417 (B) [159], 1.372 & 1.467 (C_4) [167]
Buckling parameter $\Delta_0(Å)$	0.42–0.45	0.69	0 (P), 0.55 (B) [159], 0 (C_4)
Hopping integral t(eV)	1.6	1.3	2.835 & 2.525 (C_4) [168]
Energy gap E_g (meV)	1.9	33	0
Fermi velocity $v_F(10^6 ms^{-1})$	0.65	0.62	$\sim 10^6$(B) [159]
Effective electron mass $M^*(m_0)$	0.001	0.007	–
λ_{SO}(meV)	3.9	43	–
Rashba interaction (meV)	0.7	10.7	–
ΔE(eV)	7.2	8.1	–

Source: Reproduced with permission from Matthes et al. [61].

2.2 DFT study of the optical properties

The implementation of many-body theory for the calculation of optical properties needs many-body wave-functions or Green's function and excited electronic states. It is well known in the scientific community that DFT is a ground-state theory [42–44]. So, it is outside the realm of DFT to study excitations involved in any optical calculation of a system. But the first leap towards the development of a sophisticated theory should be attributed to DFT. Also, the computational time taken by these sophisticated techniques is more than that of DFT. So, to get a firsthand idea about the optical properties of any system, DFT can be regarded as a legitimate computational tool. In fact, if we are not really concerned about the detailed optical spectra, DFT results are found to be qualitatively reasonable [45]. It should be mentioned here that the electron energy loss spectra (EELS) of pure graphene, computed by employing DFT [46], matches reasonably well with the experimental observation [47].

2.2.1 Methodology

In recent years, DFT has emerged as an important theoretical tool for predicting various physical properties related to exotic materials. This method based on supercell approach, however, takes into account the relaxation of atoms. Traditional calculation such as local density approximation (LDA) and generalized gradient approximation (GGA) underestimate the bandgaps of semiconductors and insulators. This can be, however, overcome by going beyond DFT and taking appropriately the many-body effect as is done through self-energy computation involving Green's function (G) and the screened coulomb (W) interaction or in short through GW approach.

Hybrid functionals, on the other hand, have also been proven to be another powerful technique in DFT which can produce consistent band structure comparable to experimental situation and a reliable description of charge localization often used in low-dimensional system. In particular, the screened hybrid functional due to Heyd, Scuseria and Ernzerhof (HSE) [48] has been observed a reliable one in predicting formation energy and other defect levels in semiconductors. Another hybrid functional due to Perdue, Burke and Ernzerhof (PBE) [49] has also been instrumental in addressing the electronic properties based on atomic structures.

In DFT, optical properties of any system can be calculated with the help of frequency-dependent dielectric function which is complex in nature: $\epsilon(\omega) = \epsilon_1(\omega) + i\epsilon_2(\omega)$, where $\epsilon_1(\omega)$ and $\epsilon_2(\omega)$ are, respectively, the real and imaginary part of the dielectric function as a function of energy of the incident electromagnetic (EM) wave. These two quantities are not independent of each other. Rather, they are connected by a relation which is known as the Kramers–Kronig (KK) relation [50, 51]. In any numerical simulation, the imaginary part of the dielectric function is calculated with the help of a time-dependent perturbation theory in the simple dipole approximation. In the long-wavelength limit ($q \to 0$), the imaginary part of the dielectric function is given by the following expression:

$$\epsilon_2(\omega) = \frac{2e^2\pi}{\Omega\varepsilon_0} \sum_{K,\,CB,\,VB} |\langle\psi_K^{VB}|\, \vec{u}\,.\,\vec{r}\, |\psi_K^{CB}\rangle|^2 \delta(E_K^{CB} - E_K^{VB} - \omega) \tag{2.1}$$

In the above expression (2.1), ω is the frequency of the EM radiation in energy unit. Ω represents the volume of the supercell and ϵ_0 is the free space permittivity. CB and VB represent the conduction band and the valence band (VB), respectively. $\vec{u}$ and $\vec{r}$ denote the polarization vector and position vector of EM field, respectively. The matrix element of this dot product of these two vectors is computed between the single-electron energy eigen states. Since the magnetic field effect is weaker by a factor of v/c, the transition matrix elements between the eigen states of CB and VB have been calculated only due to the electric field. Phonon contribution, local field and excitonic effects are not taken into consideration. By definition, the imaginary part of the dielectric function is positive for any polarization and frequency [6, 52–55].

As mentioned before, $\epsilon_1(\omega)$ can be obtained from $\epsilon_2(\omega)$ by KK transformation which is given by,

$$\epsilon_1(\omega) - 1 = \frac{2}{\pi} P \int_0^\infty \frac{\omega' \epsilon_2(\omega') d\omega'}{\omega'^2 - \omega^2}$$

Various sum rules involving $\epsilon_1(\omega)$ and $\epsilon_2(\omega)$ can be verified to check the consistency in the numerical computation [20].

The complex refractive index ($\widetilde{N}$) of any material is related to the complex dielectric function ($\epsilon(\omega)$) by the relation $\widetilde{N} = \sqrt{\epsilon(\omega)}$. From this relation, one can get the real as well as the imaginary part of the refractive index in the form $\widetilde{N} = n(\omega) + ik(\omega)$, where,

$$n(\omega) = \left(\frac{\sqrt{\epsilon_1^2 + \epsilon_2^2} + \epsilon_1}{2}\right)^{\frac{1}{2}}$$

$$k(\omega) = \left(\frac{\sqrt{\epsilon_1^2 + \epsilon_2^2} - \epsilon_1}{2}\right)^{\frac{1}{2}}$$

From $n(\omega)$ and $k(\omega)$, one can compute the reflectivity at normal incidence of EM wave as,

$$R(\omega) = \frac{(n-1)^2 + k^2}{(n+1)^2 + k^2}$$

The reflectivity modulation ($R_M(\omega)$) [56] can be obtained from the reflectivity data by the following expression,

$$R_M(\omega) = \frac{1}{R(\omega)} \frac{dR(\omega)}{d\omega}$$

Although $R(\omega)$ is positive for all frequencies, however, $R_M(\omega)$ can take either of sign.

The absorption coefficient can be evaluated by using the imaginary part of the refractive index as,

$$\alpha(\omega) = \frac{2k\omega}{c\hbar}$$

where c represents the speed of light in vacuum and ω is in the energy unit. The direct measure of the collective excitation is the quantity called EELS. It is given by the relation $L(\omega) = Im\left(-\frac{1}{\epsilon(\omega)}\right)$ or in terms of ϵ_1 and ϵ_2,

$$L(\omega) = \frac{\epsilon_1}{\epsilon_1^2 + \epsilon_2^2}$$

Typical energy of the plasmons of a system can be estimated by looking at the peak positions of any loss function.

The optical conductivity $\sigma(\omega)$ is related to the dielectric function via the following relation:

$$\sigma(\omega) = \sigma_1(\omega) + i\sigma_2(\omega) = \frac{-i\omega}{4\pi}(\epsilon - 1)$$

All the optical property calculations are performed in the long-wavelength limit $q \rightarrow 0$ of EM wave.

2.3 FS silicene monolayer

2.3.1 Optical properties

Because of its buckled structure, it is convenient to study the modification of electronic as well as optical properties of silicene under an external electric field. Applying external electric field is both theoretically and experimentally convenient. So, it is natural to explore the effect of external electric field on silicene sheet. But external electric field can significantly change the electronic as well as the optical properties of silicene. Kamal et al. [57] have first studied the effect of this external electric field on the electronic and the optical properties of pure silicene and hydrogenated silicene by employing DFT. It has been observed by them that the optical properties strongly depends on the direction of the incident EM field, i.e. whether it is in the plane of the silicene sheet (parallel polarization) or perpendicular to it (perpendicular polarization). The authors have found that the low-energy regime (below ~ 5 eV) and the high-energy regime (above ~ 5 eV) are mainly dominated, respectively, due to parallel and perpendicular polarization. Optical anisotropy observed in this system may be due to inherent buckling present in this 2D material system. A careful analysis of the data obtained from absorption spectra calculation has clearly indicated that there is no absorption of light below a cut-off frequency. The physical reason of this phenomena is due to the symmetry breaking between two sublattices due to the application of external electric field. The more the strength of the external electric field, the more the cut-off value has been seen. When a silicene-based device is irradiated by THz EM field which is circularly polarized, the topologically insulated spin-up and spin-down bands are respectively red- and blue-shifted [58]. While in the case of band insulator, both bands are shown to be continually blue-shifted with increasing staggered potential.

First-principles DFT calculations on hybrid S/G nanocomposite have revealed that nanocomposite exhibits stronger optical absorption in the frequency range from 0 to 15 PHz compared to silicene on graphene monolayer [59]. The interest in this composite structure originates from the fact that silicene interacts overall weakly over graphene through van der Waals interaction in such a way that their intrinsic electronic structure is restored. Besides, the interlayer interaction can interestingly induce p-type and n-type doping of silicene and graphene, respectively.

Changes in material properties with the application of strain have always been interesting in the field of material science. Mohan et al. [60] have studied the strain-dependent electronic and optical properties of silicene. They consider both asymmetric and symmetric strain (up to 20%) of equal magnitude. Within 4 eV, they have found some characteristic peaks corresponding to inter-band transitions in electronic band structure. But interestingly, it has been observed that the characteristic peaks vanish with increasing magnitude of strain. Those transitions, which occur above 2 eV, shifted to lower frequency with increasing magnitude of tensile and asymmetric strain. But with compressive strain, the transitions are blue-shifted. If we plot the real part of the dielectric function as a function of energy, then the points where it cuts the x-axis (energy) from both positive and the negative sides denote the collective excitations of electron. These excitations produce oscillations of electron's density in the system, which are known as plasma oscillations. The peaks in the EELS correspond to these oscillations. In pure silicene, two distinct EELS peaks appear. One is within 7–8 eV which is due to $\pi + \sigma$ plasmons. The other one is within 2 eV, which is due to π plasmons. But with increasing magnitude of compressive strain, the positions of the π plasmons are not affected. But the disappearance of π plasmon peaks are noticed for tensile and asymmetric strain. Optical conductivity is also very important quantity for identifying the electronic transitions within the systems. Matthes et al. [61, 62] have studied the optical conductivity of silicene. Their calculation is based on independent quasiparticle approximation. It has been revealed from the study that the low-energy peak around 2 eV is due to π–π^* transition. The high-energy peak near 5 eV is due to σ–σ^* transitions. It is due to the fact that σ bonds are much stronger than the π bonds so naturally their excitation peak should occur at higher energy.

Motivated by these intriguing features, recently Das et al. [63] have explored the optical properties of disordered silicene nanosheet. Where the pristine nanosheet has been made disordered by doping with Al and P atoms ($\sim 16\%$ for Al/P doping and $\sim 31\%$ for codoping). The concentrations of Al and P atoms have been varied from 3.12% to 15.62%. While for Al-P codoped systems, it has been varied from 6.25% to 31.25%. There are several observations in the optical properties of these doped systems. One is below 3 eV, which can be associated with π–π^* transitions. Another one is above 3 eV, which can be associated with σ–σ^* transitions. It has been found that for perpendicular polarization, two new small yet significant EELS peaks have been emerged. Absorption coefficient study reveals that the maximum value of the absorption coefficient of the doped system is higher than that of pristine system. For parallel polarization, the maximum value of the absorption coefficient is blue-shifted irrespective of the doping concentrations. But, for perpendicular polarizations, a red-shift is observed for Al and P doped systems. In Figure 2.1, we have depicted the real and the imaginary part of the refractive index for Al and P doped systems for parallel polarization. In the case of parallel polarization, the low-energy (below 10 eV) sector of the energy spectra is mainly affected. Here, again, it is observed that the signature of defects is appearing mainly in the low-energy sector of the spectrum. All of these transitions can be associated with π band transitions.

Nanostructure having closed edges are known as nanodisks. Nanodisks, which have hexagonal symmetry, are mainly derived from many benzene rings [64]. It can be experimentally synthesized by using an experimental technique

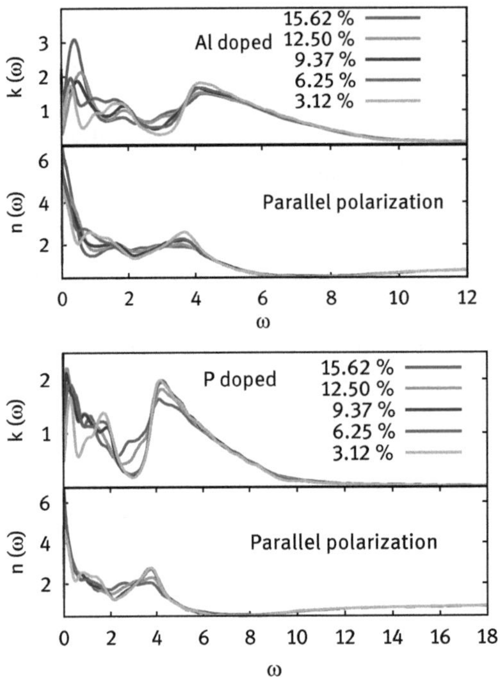

Figure 2.1: Real (*n*) and imaginary (*k*) part of the refractive index as a function of energy for (top panel) Al and (bottom panel) P doped systems for parallel polarizations.

called soft-landing mass spectrometry [65]. Regarding the optical properties of nanodisks, Sony and Shukla [66] first have studied the optical absorption spectra of various graphene nanodisks by using Pariser–Parr–Pople (PPP) model.

The magnetic and optical properties of differently shaped nanodisks have been first studied by Chowdhury et al. [67] by employing DFT. Four differently shaped nanodisks: zigzag trigonal (ZT), armchair triangular (AT), diamond shaped (DS) and bowtie shaped (BS) have been considered for the study. In Figure 2.2, we have schematically illustrated the structures of four differently shaped nanodisks. Edge atoms play a very important role in these kinds of nanodisks. Now, the percentage of the atoms at the edges of ZT, AT, DS and BS are, respectively, given by 69.2, 66.6, 62.6 and 60%. From the real part of the dielectric function data, an oscillatory behaviour up to 9 eV has been found for ZT and AT nanodisks. Beyond this energy range, the optical response becomes negligible. The static values of the real part of the dielectric function of all the nanodisks are found to be less than that of bulk Si and silicene. Different values of the dielectric function for different nanodisks can be traced due to anisotropy and different shape geometry. The comparison of optical properties of four

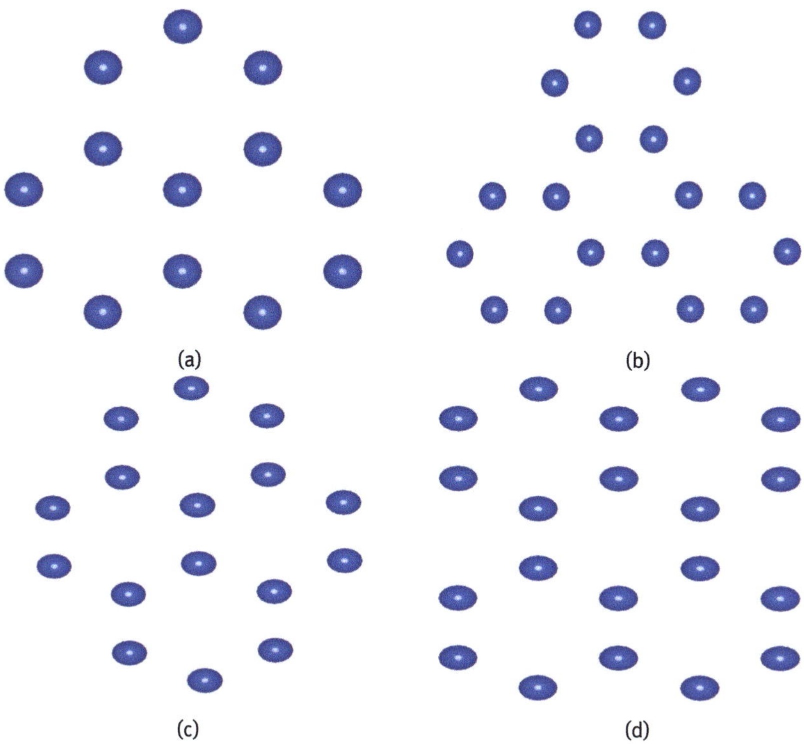

Figure 2.2: The schematic structures of four differently shaped nanodisks: (a) zigzag trigonal (ZT), (b) armchair trigonal (AT), (c) diamond shaped (DS) and (c) bowtie shaped (BS).

different nanodisks shows that DS nanodisk has the best optical response, whereas AT bears the most poor optical response. This kind of study is very important for device fabrication in nano-optoelectronic technology and material characterization techniques.

Magnetic properties of single-vacancy (SV), DV, Al and P doped silicene have already been studied by Majumdar et al. [68]. In Figure 2.3, we have illustrated the real and imaginary parts of the dielectric function for SV and DV-induced silicene nanosheet as a function of energy for perpendicular polarization. From the figure, it can be observed that the signature of vacancy is mainly appeared within 6–10 eV. Also, the effect of both SV and DV is quite similar in the optical spectrum. Due to the presence of vacancy within the sample, the optical response is poor as can be seen from the magnitude of the dielectric functions. The maximum value of the imaginary part of the dielectric function is only 0.25. In the case of the real part, it varies between 0.85 and 1.15. Wei et al. [69] in 2013 and in 2016 Zakerian and Berahman [70] have studied the optical absorption of SV-induced silicene sheet by employing both DFT

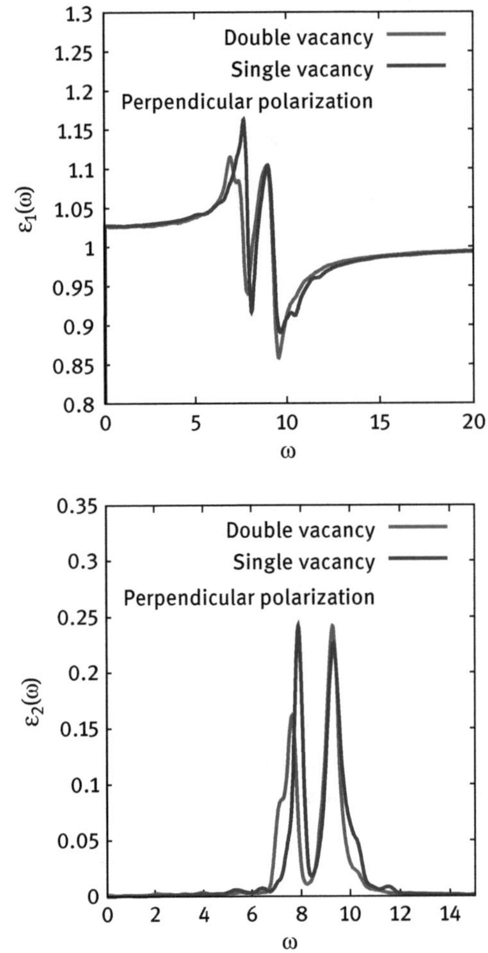

Figure 2.3: (Top panel) Real (ϵ_1) and (bottom panel) imaginary (ϵ_2) parts of the dielectric function as a function of energy for mono- and DV-induced silicene nanosheet for perpendicular polarization.

and Bethe Salpeter equation (BSE). As BSE takes into account the electron–hole interaction which DFT does not, so BSE employed data is expected to be more accurate than DFT. They have obtained two peaks in pristine silicene, one is located around 1.2 eV and the other one is around 4 eV [69]. SV-induced silicene has also shown two peaks like the pristine one, but here interestingly the two peaks are red-shifted. This happens due to the presence of dangling bond in the defected silicene. This characteristic feature can help the experimentalist to distinguish pristine silicene and defected silicene apart from Raman study.

DFT can only predict the ground-state properties of a many-body system as it is based on frozen atom approximation. But, if one can incorporate appropriate lattice dynamics and also choose thermally equilibrated configurations, then various optical properties can be computed at finite temperature which can be matched with the experiment. Now, to generate various thermally equilibrated configurations, molecular dynamics (MD) is a commonly adopted computational tool. From those configurations, one can compute the ensemble average of dielectric functions [71]. Yang and Liu [72] adopted this technique to calculate the dielectric function of monolayer silicene sheet. Apart from 0 K, they have considered two more temperatures 300 and 600 K. At each configuration, they have calculated the imaginary part of the dielectric function and then they have plotted the ensemble average of the imaginary part of the dielectric function. It has been revealed from their study that for parallel polarization, absorption peak around 1 eV energy range is enhanced consistently with increase of temperature. It has been explained through zero-energy gap and intraband transition which generally dominates the optical absorption occurring at low energy. With the increase of temperature, the amplitude of lattice vibrations is enhanced. As a result, more and more free carriers participate in the intraband transition which are thermally excited. But in the case of perpendicular polarization, due to structural disorder, the bands tend to spread out which reduce the absorption peaks. It will be interesting if these kinds of study can be done with silicene with defects.

Ye et al. [73] have studied the optical properties of periodically removed hexagonal silicon chains from silicene sheet, which is known as silicene nanomesh. While plotting the imaginary part of the dielectric function, they have observed a striking difference with that of bulk Si for both polarized and unpolarized light. Bulk Si has a threshold value of 1.1 eV below which there is no direct optical transitions occurring between the valence band maximum (VBM) and the conduction band minimum (CBM) [74]. It has been noticed that the maximum value of the imaginary part of the dielectric function lie within the visible to infrared (IR)\ part of the spectrum. The peak positions also do not noticeably change for the unpolarized and for parallely polarized light; however, the magnitude is doubled for parallely polarized light. But a strong anisotropy signal is noticed for both kinds of polarizations. The authors further commented that silicene nanomesh may pave the ways towards the fabrication of solar cell.

It is to be noted that quite a lot of theoretical works have already been done on the optical properties of silicene. But experimental works related to the exploration of optical properties are still inadequate. Sugiyama et al. [75, 76] have synthesized and also studied the optical properties of phenyl-modified (oxygen free) organosilicon nanosheet. The consequence reported of this structure is its uniform dispersion in organic solvents. The material has been synthesized by the reaction of layered polysilane [Si_6H_6] with phenyl magnesium bromide [PhMgBr]. The IR spectrum reveals the vibrations of phenyl groups at 1,150 and 1,410 cm^{-1} that correspond to the Si–Ph bond. The signature of C–C bond has been observed at 1,700–2,000 cm^{-1}. Also an asymmetric vibration of Si–H bond has been noticed at 2,100 cm^{-1}.

Photoluminescence (PL) spectrum study has also been done using 350 nm excitation. The emission spectra shows a peak at 415 nm, which is shorter than its bulk counterpart. The X-ray absorption spectrum of the silicon sheets has a peak at 268 nm, which corresponds to $L \rightarrow L$ critical point in the band structure [77]. Optical properties of alkyl-modified crystalline silicon nanosheets have been explored by Nakano et al. [78]. They have adopted different experimental techniques to characterize the system. This new structure is able to disperse uniformly in organic solvents. Fourier-transform infrared (FTIR) spectroscopy analysis produced different peaks due to bonding of different functional groups with silicon. The absence of characteristic peak at 1,600 cm^{-1} due to C–C double bond indicates that the organic molecules are covalently attached to the silicon surface. Borensztein and co-workers [79] have used an experimental technique called "surface differential reflectance spectroscopy" (SDRS) to explore the optical properties of single-layer Si on Ag (110). While analysing the SDRS signal, it has been observed that a sudden change in slope occurred at 3.83 eV. From theoretical DFT calculation, three significant peaks can be observed at 2.0, 3.8 and 4.6 eV in pristine silicene [80]. But, while silicene is deposited on Ag (110) surface, these peaks positions are supposed to change due to interlayer interface. But in the Ag surface-deposited case, no change of slope has been observed at 4.6 eV, and only a tiny effect has been seen at 2.0 eV. Through this work, the authors hope that this study will help the theoreticians in future to determine the exact atomic structure of Si layer on Ag surface.

Before ending this section, we would like to mention one more remarkable feature of the optical properties of these 2D materials. After the immediate discovery of graphene, it has been realized that the transmittance can be defined in terms of fine structure constant ($\alpha = \frac{e^2}{\hbar c}$) [82]. It was then natural to find this universal feature in other 2D materials like silicene, germanene, etc. In Figure 2.4, we compare the absorbance of graphene, silicene and germanene and confirms the universal character which is independent of the corresponding Fermi velocity and buckling [81]. It can be clearly observed that as $\omega \rightarrow 0$ the absorbance $A(0) = \pi\alpha$. It is also known [81] that as we go towards higher frequency, there is a considerable deviation of absorption spectra in the case of silicene and germanene compared to graphene. This universal feature of absorbance can be understood [62, 80, 81, 83] from the imaginary part of the dielectric function and crossing of bands linearly at the Dirac points. Here, it is noteworthy to point out that this feature is also independent of the choice of gauge, i.e. whether it is transverse or longitudinal.

2.3.2 Magnetic properties of doped FS silicene monolayer

Pristine FS monolayer silicene is non-magnetic like that of graphene [84]. But, the presence of defects like vacancy or adatoms can make silicene magnetic. This kind of metal free magnetism is now the subject of an intense research. This type of magnetism comes due to the formation of localized states caused by defects or molecular

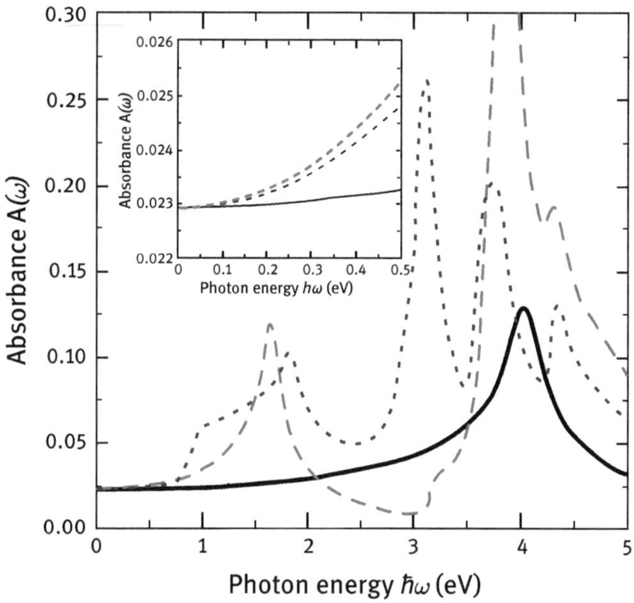

Figure 2.4: Comparison of absorbance spectra of graphene, silicene and germanene. Graphene: 0.02293, silicene: 0.02290, germanene: 0.02292. The universal behaviour is noticed for all of them, and this is independent of group IV elements, buckling and Fermi velocity (Reprinted with permission from Ref. [81]). Copyright (2012) by American Institute Physics.

adsorption. Inducing magnetism into non-magnetic nanostructure is technologically very important for making quantum information and spintronic devices. Below, we will review some of the recent work based on magnetic properties of monolayer FS silicene sheet.

The phenomenon of ferromagnetism observed at reasonably high temperatures in some compounds which do not contain any atoms with open d or f shells is known as d^0 ferromagnetism [85, 86] e.g. in HfO_2 [87], ZnO [88–90] with vacancies and so on. In most cases, ferromagnetism does not appear in the bulk when the system is pure. Thus, lattice defect is necessary for the occurrence of d^0 ferromagnetism. They may be point defects – atomic vacancies or interstitials – induced by irradiation or thermal treatments (annealing) or by lattice mismatch, grain boundaries and dislocations can also induce this d^0 ferromagnetism. In other words, the defect states can give rise to an appreciable amount of magnetic moment in connection with the molecular orbitals, which are localized very close to the defect site [86]. The utility of d^0 ferromagnetism lies in the fact that large magnetization like that of Fe, Co and Ni can be achieved in a nanomaterial having an active defect. Thus, one can visualize magnetism in non-magnetic semiconducting matrices with some appropriate impurities/voids of non-magnetic atoms. Hence, d^0 magnetism can take an important key role in designing novel materials in spintronics at room temperature. The nanoparticle surface area in 2D systems seems to be an integral part in triggering this d^0 ferromagnetism in contrast to the bulk one. Here, it is to be noted that, in some alkaline-earth metal nitrides XN (X=Ca, Sr, Ba), half-metallic d^0 ferromagnetism has been predicted without any defects [91, 92]. Recently, we have studied the d^0 ferromagnetic properties of a new kind of material which is the hybrid of graphene and silicene [93–95]. From our study, it has been revealed that, due to a single silicon atom vacancy, the system possesses a magnetic moment of $4\,\mu_B$. However, for a single carbon atom vacancy, the ground state has been found to be non-magnetic. From the charge density analysis, we have noticed that, because of the silicon atom vacancy, the three undercoordinated carbon atoms are unable to form any covalent bond, which makes the system magnetic. However, for carbon void system, the undercoordinated silicon atoms form covalent bond, making the system non-magnetic [96].

Hydrogenation and halogenation is one of the viable route to induce magnetism in silicene. Zheng et al. [98] have employed the first-principle calculation to investigate the magnetic properties of silicene sheet adsorbed with H and Br atoms. It has been found that when the silicene sheet is fully saturated with H and Br atoms, the ground state is found to be non-magnetic. But when the sheet is half-saturated from one side, it shows ferromagnetic property. It is due to localized and unpaired electrons of the unsaturated Si atoms. Total energy calculation reveals that half-hydrogenated silicene exhibits ferromagnetic order [99], while half-brominated one exhibits antiferromagnetic ordering. Paszkowska and Krawiec [100] have explored the stability of magnetism in hydrogenated silicene under the influence of strain, charge doping and external electric field. It has been found that the magnetism is present in strained hydrogenated silicene unless a structural phase transition occurs. As long as the hydrogenated silicene maintains its LB structure, strain does not influence the magnetic property of the system. However, when an external electric field is applied, then it has been observed that the magnetic ground state is still maintained. But interestingly, the magnetism disappeared for both electron and hole doping. Zhang et al. [101] have investigated the magnetic properties of different hydrogenated conformer of silicene sheets. It has been noticed that half-hydrogenated chair-like conformer shows magnetic ground state. This is because of the p_z electrons of the unhydrogenated Si atoms which are unpaired and strongly localized to induce magnetism in the system. On the other hand, fully saturated sample is non-magnetic because of the absence of these p_z electrons due to bonding with H atoms. Zhang and Yan [102] also reported similar kind of results for fully and half-hydrogenated silicene sheet giving similar physical explanation. Adsorption of adatoms on silicene sheet is also found to be effective in tailoring magnetism in silicene sheet. Ju et al. [97, 103] have explored the magnetic properties of silicene nanosheet with changing adsorption coverage of H, F and C atoms. In Figure 2.5, the total magnetic moment of C and H adsorbed systems have been depicted with different coverage density. It has been

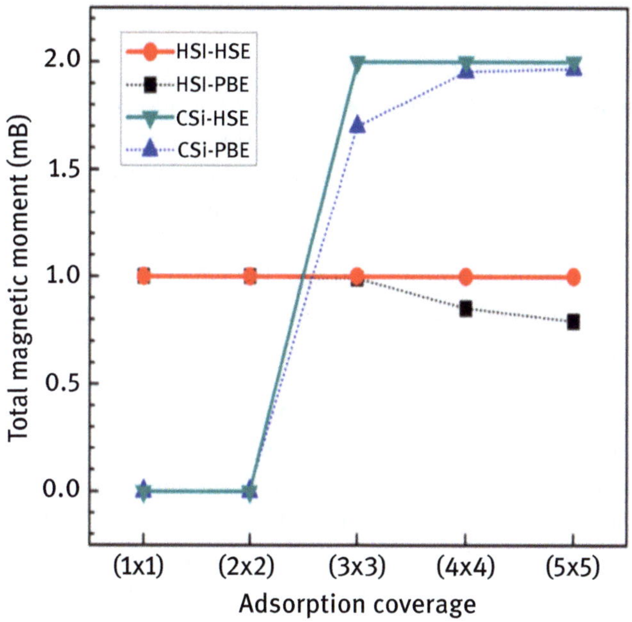

Figure 2.5: The total magnetic moments of the spin-polarized systems as a function of adsorption concentration (Reprinted with permission from Ref. [97]). Copyright (2016) by Elsevier.

reported that adsorption of F atoms makes the sample non-magnetic; however, H and C atoms induce magnetism within the sample for different adatom coverage. It has been observed that the total magnetic moment of the system is same (1 μ_B) for different H coverage density. This value, however, remains constant and not affected by H concentration. In the case of C atoms adsorbed systems, among five different configurations, only for three configurations with low coverage density, magnetic ground states have been reported with the value of 2 μ_B.

Inducing magnetism in a non-magnetic material by transition metal (TM) atoms [104] is a conventional process due to their half filled *d* and *f* orbitals. They show diverse structural, electronic and magnetic properties. If TMs are adsorbed, depending on the type of adatom and atomic radius, the system can exhibit metal, half-metal and semiconducting behaviour. Motivated by the diverse properties of TM atoms, it would always be interesting to study the effect of TM atoms on silicene surface which is more reactive than that of graphene. Le et al. [105] have studied the adsorption of 3d-TM atoms on silicene using DFT+U formalism. They have considered four types of TM atoms: Cr, Mn, Fe and Co. According to the data obtained by them, the total magnetic moments of the considered (2 × 2) supercell consisting of four Fe atoms is 2.21 μ_B/cell. Further, it has been reported that each Fe atom produces a ferromagnetic moment of 3.07 μ_B, whereas each Si atom produces an antiferromagnetic moment of 0.43 μ_B. Similarly, when Co atom is used in place of Fe, non-magnetic ground state is obtained. When the other two adatoms have been used, much higher ferromagnetic moment of magnitude 4.01 and 5.18 μ_B/cell, respectively, for Cr and Mn have been obtained. Strain-dependent magnetic properties of silicene doped with Cr and Fe have been explored by Zheng et al. [106]. They have mainly considered isotropic uniaxial tensile strain to study strain-tunable magnetism. The unstrained system possesses around 2 μ_B when it is doped with Cr atom. When a small isotropic tensile strain (~ 3.5 %) is applied, the magnetic moment suddenly jumps to around 4 μ_B. Similar phenomena is observed in the case of Fe atom doped system. Where in the unstrained case, its magnetic moment is around ~ 0.4μ_B. But when a small tensile strain (~ 2 %) is applied, it reaches a very high magnetic state of magnetic moment of 4 μ_B. A hysteresis is also observed for both kinds of doped system specially for Fe doped systems. Sahin and Peeters [107] have also studied the adsorption of 3d-TM atoms along with alkali and alkaline-earth metal atoms. It has been demonstrated by them that alkali and alkaline-earth metal atoms are ineffective in inducing any magnetic moment in the adsorbed systems. However, significant amounts of magnetic moment have been noticed for 3d-TM atoms. Their study has revealed that the amounts of magnetic moment per unit cell for Ti, V, Cr, Mn, Fe and Co are, respectively, 2, 2.7, 4, 3, 2 and 1 μ_B/cell. Ozcelik and Ciraci [108] have also explored the adsorption of Si, H and Ti atoms on silicene by employing DFT. From the obtained data, they have noticed the magnetic ground state of silicene. In Table 2.2, we have

Table 2.2: Table for magnetic moment (in μ_B) for different adsorbed (A) and doped (D) atoms induced monolayer silicene systems. The corresponding references are given in the third column.

Atoms	Magnetic moment (μ_B)	Reference
Cr (A)	4.01/cell	[105]
Mn (A)	5.08/cell	[105]
Fe (A)	2.21/cell	[105]
Co (A)	0.0/cell	[105]
Cr (D) (unstrained)	2.0	[106]
Cr (D) (strain ~ 3.5 %)	4.0	[106]
Fe (D) (unstrained)	0.4	[106]
Fe (D) (strain ~ 2.0 %)	4.0	[106]
Ti (D)	2.0	[107]
V (D)	2.27	[107]
Cr (D)	4.0	[107]
Mn (D)	3.0	[107]
Fe (D)	2.0	[107]
Co (D)	1.0	[107]
Si (A)	2.0	[108]
H (A)	1.0	[108]
Ti (A)	2.0	[108]

provided the magnetic moment of different doped and adsorbed atoms induced monolayer silicene systems.

Vacancy-induced magnetism has also been one of the interesting topic in condensed matter and material physics. Here also the absence of *d* orbital is noteworthy. SV, DV and triple-vacancy (TV)-induced buckled silicene have been reported to be non-magnetic [109]. However, for SV-induced graphene, ferromagnetic ground state has been found to be energetically favourable [109]. In SV-induced graphene, there are three atoms which are twofold coordinated, and each of them has a dangling bond. MD simulation reveals that, upon bond reconstruction, three of them reconstruct in an asymmetric way. Among three, two of them form a C–C bond to make each of them threefold coordinated. The remaining atom, however, is unable to make any bond contributing a net magnetic moment in the system. While for SV-induced silicene, the three twofold undercoordinated atoms come close to each other to form three Si–Si bond. This makes the whole system non-magnetic. In the case of DV-induced graphene and silicene, there are four twofold coordinated atoms surrounding the vacancy. During bond reconstruction, four dangling sp^2 bonds are saturated in pairs to form two C–C or Si–Si bonds. In this way, both the systems possess non-magnetic ground state. For TV, the ground state of graphene is found to be magnetic, while for silicene it is non-magnetic. We have depicted in Figure 2.6 the magnetic properties of Al and P doped disordered silicene systems. It can be seen from Figure 2.6 that for Al doped systems, when the defect concentration reaches 9.37% and 15.62%, the net magnetic moment of the system reaches nearly 1 μ_B. While for P doped systems, the magnetic moment is respectively 0.27, 1 and 0.42 μ_B for 6.25, 9.37

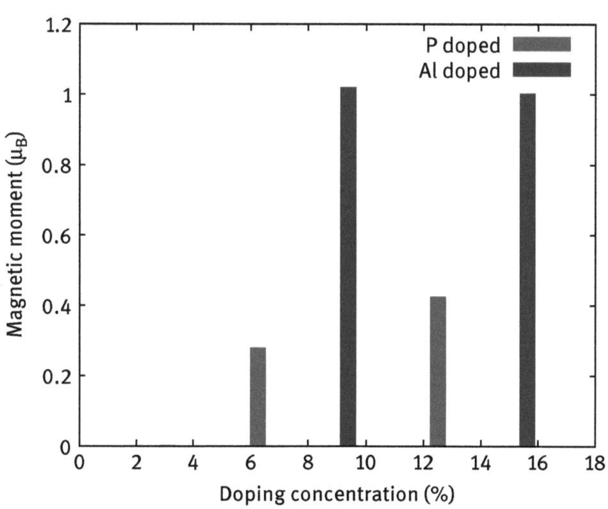

Figure 2.6: Magnetic moment (in μ_B) as a function of defect concentrations for (left panel) Al doped and (right panel) P doped silicene sheet.

and 12.50% doping concentrations. It will be interesting to study in future the stability and tuning of this magnetism under the combined influence of strain and external electric field. When TM atoms are embedded in vacancy-induced silicene sheet, magnetism can be observed [110]. In Figure 2.7, we have illustrated the magnetic moments of both SV- and DV-induced silicene sheet embedded with TM atoms. From the figure, it is clear that Sc and Ti embedded systems are non-magnetic. It is because, in the valence shell Sc has three electrons, which is lesser than that of Si, so the *d* orbital gets saturated by forming three Sc–Si bonds in SV and four Sc–Si bonds in DV and possesses non-magnetic ground state. Ti has four valence electrons, which is the same as Si, so non-magnetic states have been observed both for single and for double vacancy. However, for V, Cr, Mn, Fe and Co, magnetic states are observed for both kinds of vacancies. Again for Ni, Cu and Zn, the systems possess no net magnetic moment because of the saturation of all the 3d electrons. To study the magnetic property further, SV-induced systems have been further doped with N and C atoms before embedding with TM atoms. The results are depicted in Figure 2.8. Here also, Sc, Ti, Cu and Zn produce non-magnetic ground state. But here interestingly Ni, which earlier produces non-magnetic ground state, now produces magnetic moment in the system. Ghosh et al. [111] have explored the electronic and magnetic properties of silicene with extended line defects. The defect induces 0.2 μ_B amount of magnetic moment to the system. This defect-induced system also exhibits spin localization, whose origin can be explained on the basis of buckling in silicene. The intrinsic buckling in silicene localizes the $3p_z$ electrons more strongly than that in graphene.

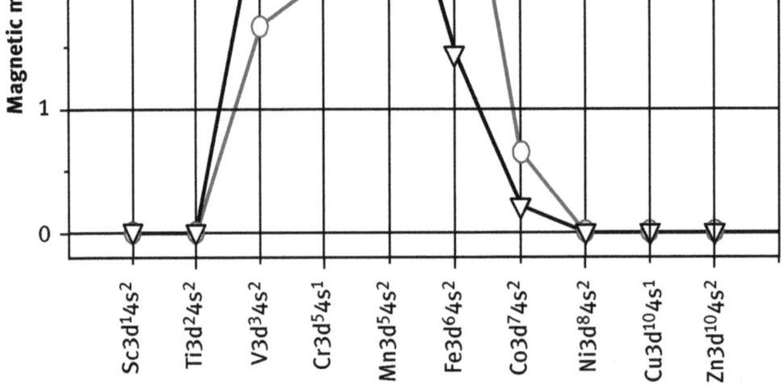

Figure 2.7: Magnetic moments of silicene with embedded TM atoms in SV (red) and DV (black) defects (Reprinted with permission from Sun et al. [110]. Copyright @ American Institute of Physics (2015)).

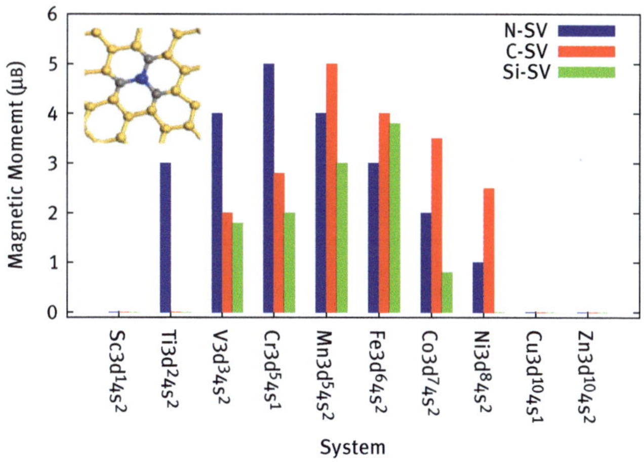

Figure 2.8: Bar diagram of magnetic moments of TM atoms embedded into N or C doped SV defect in silicene. The green bars indicate the magnetic moments of transition metal atoms embedded into SV without doping for comparison. The inset shows the configurations of TM atoms in C-doped SV (Reprinted with permission from Sun et al. [110]. Copyright @ American Institute of Physics (2015)).

The use of superhalogen ($MnCl_3$) over conventional halogen to tune the electronic as well as magnetic properties of silicene has been explored extensively by Zhao et al. [112]. The advantage of $MnCl_3$ over conventional halogen is that it has even larger electron affinity than that of Cl atom. Isolated $MnCl_3$ is intrinsically magnetic. So, it would be interesting to study the effect of $MnCl_3$ on silicene; both of them have been experimentally realized in laboratory. Different adsorption sites have been considered. For the convenience of the readers, in Figure 2.9, we have schematically illustrated different adsorption sites for monolayer silicene. It has been exhibited that $MnCl_3$ prefers to occupy the hollow site (HS) of silicene. But its magnetic behaviour is quite complex. The magnetic moment of isolated $MnCl_3$ is 4 μ_B. But when one $MnCl_3$ molecule is adsorbed into a supercell, the total moment of the supercell remains 4 μ_B. This magnetic moment changes to 3 and 5 μ_B if a Cl atom is introduced on the same side or on the opposite side of $MnCl_3$. However, two $MnCl_3$ superhalogens adsorbed onto the same side result in a total magnetic moment of 8 μ_B per supercell. It will be interesting to study the effect of this superhalogen on an anisotropic and isotropic strained silicene.

The optical properties of four differently shaped silicene nanodisks [67] have been described in the previous section. It has been found that among the four nanodisks, only ZT silicene nanodisk is magnetic in nature having $\sim 4\,\mu_B$ of magnetic moment. Earlier studies [113–115] have also found similar results for graphene and silicene nanodisks. The origin of this magnetism in ZT nanodisk can be traced back

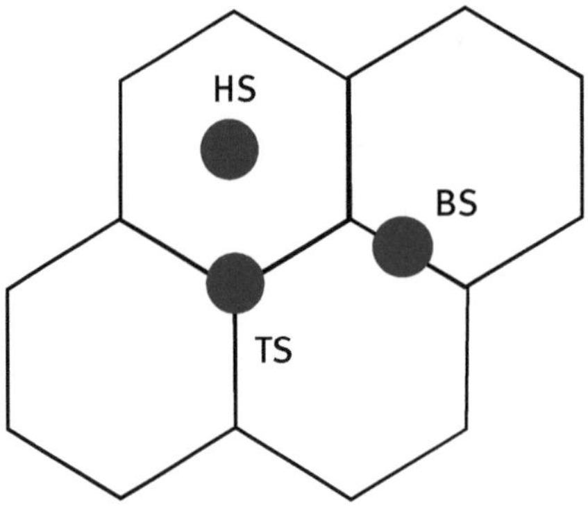

Figure 2.9: Top view of the schematic representation of adatom adsorbed silicene on different adsorption sites: (i) on hollow site (HS), (ii) on bridge site (BS) and (iii) on top site (TS) (Reprinted with permission from Nath et al. [53]. Copyright @ Elsevier (2015)).

due to the occurrence of zero-energy states [113–115] which are nothing but singly occupied molecular orbital [116]. It is now believed that Coulomb exchange interaction plays a very important role to align the spins at the edge atoms in ZT nanodisk which gives rise to magnetism. AT nanodisk also has zero-energy states, but at the edge it has even number of atoms which force it to be non-magnetic. Due to this large amount of magnetic moment, one may think ZT nanodisk as a potential candidate for future spintronic devices.

2.4 Elemental structure and synthesis of FS germanene

Germanene follows the hexagonal honeycomb-like atomic arrangement of graphene with two atoms per unit cell. As mentioned earlier, SOC plays an important role in various properties of these 2D materials. Germanene possesses a very high value of SOC (~ 46.3 meV) in comparison to that of graphene and silicene [117], which causes a mixture of sp^2 and sp^3 hybridization more favourable in germanene, despite only sp^2 hybridization, in the case of graphene [29, 118]. As a result, a finite value of buckling parameter has been introduced in planar geometry of graphene in the formation of stable FS germanene. Cahangirov et al. have adopted finite-temperature MD calculations to show that LB honeycomb structures of Ge can be stable [118]. Nijamudheen et al. have also confirmed about the fact that Ge–Ge bond is unstable in a planar geometry [119]. Reported value of this buckling parameter, $\Delta = 0.64–0.69$ Å for germanene [118, 120, 121]. This higher value of buckling plays a crucial role in different electronic and chemical properties of germanene. Deng et al., through a DFT-based calculation, have reported that the bond length increases and bond angle decreases from graphene → silicene → germanene [122]. Predicted values of bond angle and Ge–Ge bond length are ~ 112.50° and 2.44 Å for 2D germanene [121–123].

By analysing phonon modes, Sahin et al. have confirmed that mono-layer honeycomb structures of group IV elements and III–V binary compounds, including germanene, always possess positive phonon frequency (ω) [93] in ω–k diagram. Moreover, estimated carrier velocity around Dirac point is 0.95×10^6 m/s for buckled graphene-like Ge, which is greater than that of graphene and silicene [29]. Ye et al. have investigated the effect of buckled structure on intrinsic carrier mobility (ICM) of germanene by using DFT. They have concluded that ICM of germanene can even be greater than that of graphene and silicene [124]. High values of carrier mobility and ICM of germanene enforce this material more acceptable in semiconductor nanodevice industry-based applications beyond graphene. Crystal structure of germanene can be analysed crucially by identifying experimental Raman lines. Raman modes of FS germanene have been studied theoretically by Scalise et al. [125], and they have established main Raman active mode at around 290 cm^{-1}.

Germanene has been synthesized by Davila et al. [126], for the first time, through molecular beam epitaxy (MBE) using a gold (111) surface as substrate [126]. In the same year, almost in parallel, another group, Li et al., have successfully fabricated graphene-like germanene sheet on a Pt(111) surface. They have demonstrated clearly that the germanene grown on this surface was a 2D continuous layer with a buckled configuration [127]. Moreover, in order to synthesize germanene using a particular substrate, it is essential to gain sufficient knowledge about interaction of FS germanene with that substrate. Liu et al. in a work [128] have investigated about the interaction between oxygen (O_2) atmosphere and FS germanene using DFT and have come to a conclusion that germanene is more stable than silicene in O_2 atmosphere. In another very recent experimental study, Davila et al. have explored an experimental evidence of Dirac cone in few layer epitaxial germanene, synthesized on a gold template [129]. Despite the fact that it is not possible to synthesize FS germanene experimentally till date, several exceptional theoretical and preliminary experimental investigations have inspired scientists to explore more details about germanene for applications in future-generation 2D nanoelectronics.

2.4.1 Electronic and magnetic properties of FS germanene

Germanene exhibits similar electronic band structure like graphene with linear band dispersion relation and zero bandgap near Dirac K points if effect of SOC is not taken into account [118, 130, 131].

However, the inclusion of SOC introduces a non-zero value of bandgap at Dirac K point along with Dirac cone-like signatures, which again ensures about more fascinating properties and hence effective applications of germanene in optoelectronics, photo-voltaics, etc. Liu et al. have predicted [27] through a systematic first-principles investigations that an appreciable gap of 4 meV and 23.9 meV can be opened at Dirac points for planar and LB germanium, respectively, by SOC. These features are clearly depicted in Figure 2.10(a) and (b).

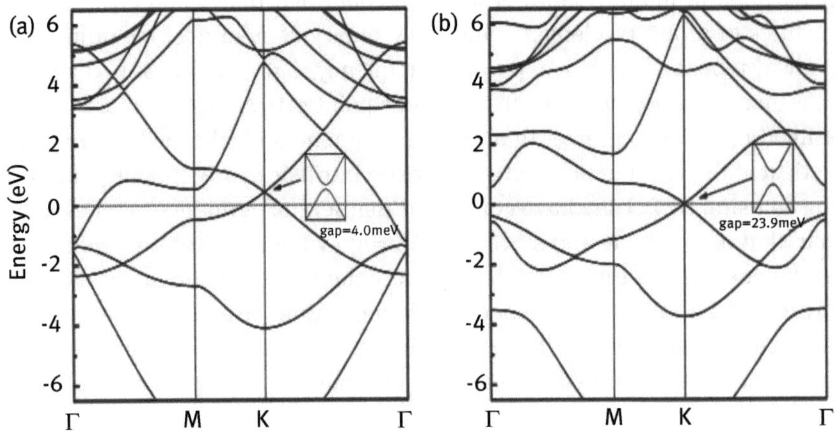

Figure 2.10: Relativistic band structure of germanium with honeycomb structure. (Left panel) the relativistic band structure of germanium with planar and (right panel) low-buckled honeycomb structure. Inset: Gap induced by SOC at Dirac K point (zooming).

Large amount of SOC indicates that germanene can exhibit QSHE in an experimentally accessible temperature regime, even at near room temperature [132]. Zhang et al. have studied the electronic properties of germanene sheets that are found on Ge_2Pt crystals after deposition of Pt on Ge(110) substrates. They have observed a V-shaped DOS which is indicative for a 2D Dirac system [133]. Besides, Behera et al. have also reported that energy dispersion relation near Dirac K point is linear for germanene by first-principles calculations [134]. Houssa et al. [135] have predicted germanene as metallic in nature with a low DOS at Fermi energy (E_F) by DFT calculations using LDA. Moreover, Lebegue et al. have also strongly confirmed germanene as a poor metal possessing small non-zero DOS at E_F by using both LDA and GGA [136]. Walhout et al. by an experimental study investigated about temperature dependence of the DOS of germanene grown on Ge/Pt crystals. They have successfully found V-shaped DOS, the hallmark of a 2D Dirac system [137]. They have explored about the fact that substrate effect can modify the original DOS of FS germanene significantly. Li et al. through a very recent DFT calculation [138] have proposed a new approach to fabricate germanene via dehydrogenating H-Ge, having the same Dirac electronic properties.

Hydrogenated silicene and germanene, termed as silicane and germanane respectively, also possess very interesting electronic properties. M. Houssa et al. [139] have predicted that in both silicane and germanane, there is a finite band opening in the band structure. The type of gap in silicane is direct or indirect depending on its atomic configuration (chairlike or boatlike). But in the case of germanene, there is always a direct bandgap opening of about 3.2 eV independent of atomic configuration which makes this material potentially interesting for application in optoelectronic devices (shown in Figure 2.11(a), (b), (c) and (d)).

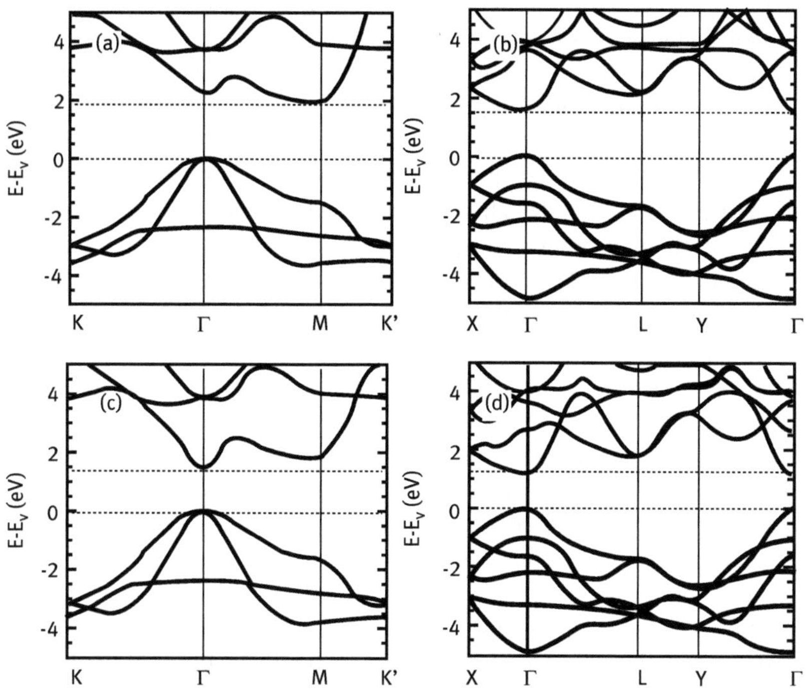

Figure 2.11: Band structures, calculated using LDA functional, (a) chairlike silicane, (b) boatlike silicane, (c) chairlike germanane and (d) boatlike germanane. Reference zero-energy level is set to the top of the valence band.

Rupp et al. [140] have reported about modifications in stability and electronic properties of germanene in the presence of impurity atoms using DFT. They have observed that the adsorption of one hydrogen (H) atom by boron (B) or nitrogen (N) impurities leads to p- and n-type semiconducting properties, respectively. Germanene, like all other 2D honeycomb structures of III–V binary compounds, is non-magnetic in nature [141]. Electronic and magnetic properties of germanene can be tuned by doping or adsorption of foreign elements in pristine system [141] or introducing vacancy or applying some strain engineering. Li et al. have proposed [142] an efficient technique of band opening by nanopatterning germanene into super-lattices using DFT. They have indicated that the broken sublattice symmetry in nanopattterning germanene is the reason for opening of bandgap. Recently, Liang et al. have explored [143] about the fact that surface functionalization and strain will modify electronic and magnetic properties of hydrogenated, fluorinated and chlorinated germanene by employing DFT. They have highlighted that fluorinated germanenes are energetically more stable than hydrogenated and chlorinated germanenes because fluorine atoms possess stronger electronegativity. Pang et al. [144] have studied the effect of alkali metal (AM) atoms adsorbed in germanene and indicated that it is possible to tailor both the gap and the concentration of charge carries of AM/germanene systems by controlling the coverage of AM. From their analysis, it is also evident that AM/germanene could be of great interest in nanoindustry, like FET applications due to strong binding of AM atoms with germanene and exceptional interesting properties of AM/germanene systems. In another work, Pang et al. have analysed [145] about the structural, electronic and magnetic properties of 3d-TM adsorbed germanene. It was concluded that in most TM/germanene structures TM–Ge bonds exhibit mostly covalent chemical bonding character. Electronic and thermal properties of germanene has been studied recently by Zaveh et al. [146] employing DFT and density functional perturbation theory (DFPT). They have observed that specific heat at constant volume (C_V) varies with temperature (T) as T^2 at low temperature from their calculations, which is consistent with the general argument that $C_V \propto \omega^{d/s}$, for excitations obeying the dispersion $\omega \propto k^s$ in an arbitrary spatial dimension d. Li et al. have examined systematically [147] about alkali, alkali-earth, group III and TM adatom adsorption on bare germanene sheet employing first-principles theory. They have elucidated the fact that interaction between metal adatoms and germanene is quite stronger than that of graphene, due to large buckling in germanene. Moreover, Kaloni has predicted that HS is energetically more favourable for 3d-TM adsorption and TM adsorbed germanenes possess magnetic moment ranges from 0.97 to 4.95 μ_B [120]. So, it is possible to induce and tailor magnetism in non-magnetic germanene by incorporation of suitable adatoms in pristine system. Xia et al. have predicted by a DFT-based study on gas adsorption of germanene that germanene can be used as a gas sensing element efficiently [148]. So one can think to utilize germanene in the implementation of sensor device-based technology. Ni et al. have explored [23] that semi-metallic LB germanene is more suitable than semi-metallic planar graphene for inducing a finite value bandgap. This suggests that germanene is more efficient than graphene for practical applications. Ozcelik et al. have predicted some new extraordinary phases in Ge adatom adsorbed on germanene by employing first-principles calculations [149]. They have revealed that, through an exothermic and spontaneous process, Ge adatom constructs a dumbbell (DB) structure on germanene. These stable DB-based phases exhibit unique electronic and magnetic properties, which can be modified by controlling the coverage of DBs. For example, at high coverage metallic state is always maintained by germanene + DB phases , but semiconductor state can also be achieved by changing different parameters like DB–DB distances or size of unit cell. Besides, Gurel et al. have investigated about the modifications of electronic, magnetic and chemical properties of germanene by charging and applying perpendicular electric field. It can be concluded from their study that charging maintains hexagonal honeycomb symmetry in germanene, but the symmetry is broken by electric field. For this, a band splitting occurs, and the value of bandgap depends linearly on applied electric field [150]. Ye et al. have concluded [151] from an ab-initio study that, it is possible to open a bandgap ~0.02–0.31 eV at the Dirac point in germanene, by the adsorption of AM atoms. Interestingly, they have also confirmed by effective mass calculation that carrier mobility in germanene remains unaffected by AM adsorption which is necessary for device application. Band structure of FS germanene can be reconstructed by inducing suitable n- or p-type adatoms like arsenic (As) and gallium (Ga) [152]. Doping has been incorporated at same or different sublattice positions of same hexagonal unit cell, which is at equivalent sites or non-equivalent sites, respectively. Different structures including pristine have shown in Figure 2.12(a)– (i), namely, pristine, S1, S2, S3, S4, S5, S6, S7 and S8, respectively.

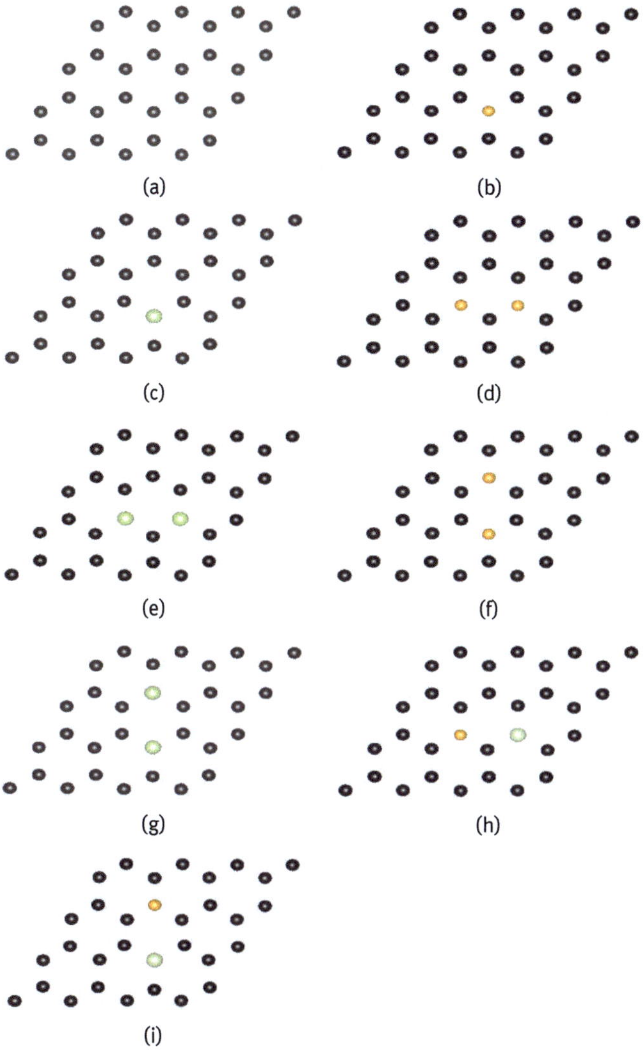

Figure 2.12: Structures (a) pristine, (b) S1, (c) S2, (d) S3, (e) S4, (f) S5, (g) S6, (h) S7, (i) S8. Largest and light-green coloured atoms are Ga, medium and black coloured atoms are Ge, smallest and deep-yellow coloured atoms are As.

It is predicted from DFT calculations that semi-metallic germanene is transformed to metallic nature by incorporating single or double doping of As and Ga, whereas semi-metallic property is preserved for AsGa codoped configurations, which can be justified by observing the position of Fermi level (E_F), number of CB and VB which cross E_F and analysing DOS. Thus, electronic and magnetic properties of FS germanene can be tailored by different ways like doping, adsorption, strain engineering and applying an external uniform electric field.

2.4.2 Optical properties of FS germanene

Now, before going for the analysis of optical properties of germanene, we would like to review the optical properties of bulk Ge briefly. Values of static dielectric constant $\epsilon_1(0)$ and IR refractive index for bulk Ge have been reported as ~ 15.9 [153] and ~ 4 [154], respectively. Wei et al. have investigated that many-body effect plays a crucial role in the different properties of germanene. From absorption spectra analysis, they have concluded that, for germanene, $\pi \rightarrow \pi^*$ resonant excitation appears at an energy value 1.10 eV with a binding energy 0.82 eV. They have further indicated that excitonic effect in 2D germanene is much stronger than that of graphene [155]. Bechstedt et al. have explored about IR absorbance of germanene by GGA–DFT method. They have demonstrated that value of IR absorbance for germanene is 0.02292, which is nearly equal to $\pi\alpha = 0.022925$, where $\alpha = 1/137.036$, the Sommerfeld fine structure constant [81]. Interestingly, it has also been concluded that this value of IR absorbance is a universal characteristic feature of studied group IV 2D materials, independent of materials and values of buckling [61, 80] . Optical conductivity of 2D germanene sheet has been studied by Matthes et al. using DFT. They have reported that in optical conductivity, most intensity peaks with phonon energy can be described by three damped harmonic oscillators in the region 0 to 10 eV [62]. Optical properties of germanene have been investigated by Pulci et al. using many-body perturbation theory [83]. Exciton binding energy and oscillator strengths in germanene have been found to be very strong, which make it a novel promising material in 2D nanoindustry. Imaginary part of dielectric constant ($\epsilon_2(\omega)$) for silicene and germanene are shown in Figure 2.13. It can be depicted from this figure that, π plasmon peak appears at 1.2 eV and 1.4 eV and $\pi + \sigma$ plasmon peak appears at 3.93 eV and 3.1 eV for silicene and germanene, respectively [156, 157].

Optical properties of FS germanene can also be modified by incorporating doping of foreign elements significantly. It has been elucidated that the value of universal IR absorbance can be enhanced or reduced than pristine germanene by incorporating suitable combinations of doping elements, site of doping and concentration of doping [152]. Different optical properties in terms of the real part of dielectric constant have been investigated ($\epsilon_1(\omega)$) using first-principles DFT methodology. The optical properties of AsGa codoped structures were also studied schematically.

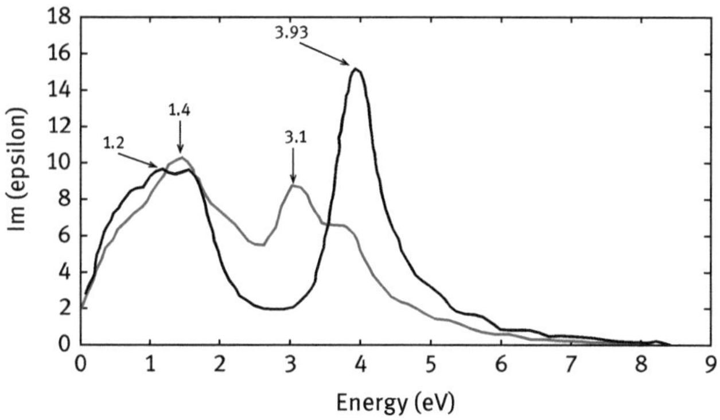

Figure 2.13: Imaginary part of dielectric constant for FS 2D materials silicene (black-dashed) and germanene (red-dotted) calculated using DFT.

To analyse the optical properties, the electric field is applied for perpendicular polarization, that is along the axis (Z axis) perpendicular to the plane of germanene sheet. Value of $\epsilon_1(0)$ is 1.61 and there are no prominent modifications in $\epsilon_1(0)$ due to doping as depicted from Figure 2.14(a)–(c). Plasma frequency (ω_P) is defined by the energy position where $\epsilon_1(\omega)$ is equal to zero. Number of ω_P for pristine layer is six, whereas the same is reduced to four, three, two and one in case of structures S4, S7, S6 and S8, respectively. As doped structures exhibit red-shifting nature of peaks and Ga doped structures possess blue-shifting nature of peaks as doping concentration is increasing, whereas peak positions remain almost unchanged in the case of codoped structures with respect to pristine layer. Moreover, it is also possible to tune other optical properties like EELS, optical absorption spectra, refractive index, reflectivity and optical conductivity by doping of impurity atoms in germanene. So it is noteworthy to conclude that the optical properties of germanene can be tailored in the presence of foreign elements, which may be helpful in the application of 2D optoelectronic industry, based on beyond graphene materials.

2.5 Structural properties of TG sheet

In their study [39], Enyashin et al. have used the DFT-based tight-binding method (DFTB) to successfully accomplish the optimizations of all the 12 structures by minimizing the total energy and forces between atoms up to reasonable satisfactory accuracy. Among those configurations, a structure with equal number of squares and octagons was previously studied by Zhu et al. [158]. The calculation [158] demonstrated that this structure with tetragonal symmetry retains its planar form even after geometrical relaxation. The relaxed network consists of a single type of sp^2 bond with bond length 1.429 Å and lattice constant 3.47 Å. In addition, they also explored that the calculated total energy per atom of this structure is greater than

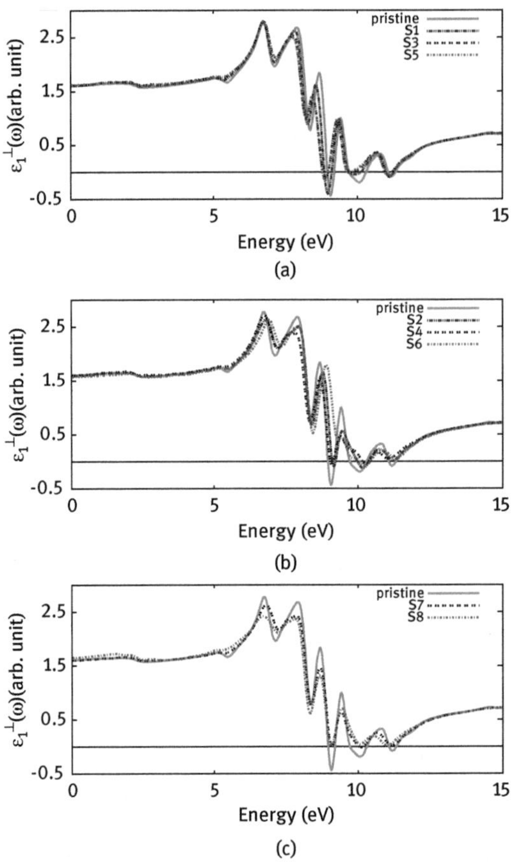

Figure 2.14: Real part of dielectric constant for perpendicular polarization. (a) As doped structures, (b) Ga doped structures, (c) AsGa codoped structures.

that of graphene. So, this structure known as TG is meta-stable compared to graphene. Inspired by these studies, Liu et al. [159] focussed their interest on this structure after proper relaxation and indicated that, it can be described by planar group p4mm with lattice constant 3.42 Å, duly supported by previous data. The calculated formation energy (−8.73 eV/atom) under GGA approximation have confirmed TG as a thermodynamically meta-stable structure. In addition, thermodynamic stability of this structure is greater than any other graphene allotropes, roughly below a certain high temperature (900 K). Besides planar one, Liu et al. [159] have also introduced another buckled network with similar tetragonal symmetry. They have named it buckled TG and described that this energetically metastable structure with formation energy −8.41 eV/atom is potentially more sound than planar one. The numerical calculations have claimed that the lattice constant and an equilibrium height difference (Δz) between two nearest-neighbour tetrarings of buckled TG is 4.48 Å and 0.55 Å, respectively. However, the structural stability of buckled TG has been challenged by Kim et al. [160]. As they have showed that the buckling of the fully relaxed TG is 0.0002 Å which is 250 times smaller than the previous reported value by Liu et al. [159]. Therefore, it is not possible to distinguish a fully relaxed buckled TG with the planar one. In a study related to non-hexagonal 2D material-based nanosensors, Liu et al. [161] also supported the fact that relaxed buckled TG transformed into the planar TG. In reply, Liu et al. [162] have not only conceded the fact but also explains the reason as planar TG has lower formation energy than the buckled one. However, they proclaimed that it is inappropriate to verify the stability of buckled TG using simple bonding arguments. Because the formation energy is necessary but not the sufficient criterion for determination of structural stability. They also pointed out that buckled TG is stable at high temperature (above 940 K). This argument is however not new, as Cahangirov et al. [118] already illustrated that a well-studied structure LB silicene is stable despite a minimum binding energy. Nevertheless, among these two structures, planar TG is studied more, not only as a 2D sheet but also as quasi-1D ribbons and 0D clusters.

2.5.1 Electronic properties of pristine and functionalized TG sheet

The electronic properties of planar TG was first reported by Enyashin et al. [39]. Their calculated band structure has revealed that E_F lies below the top of the VB, which confirms its metallic nature. This metallic nature is also supported by its appreciable (non-zero finite) DOS at E_F. The distribution of electronic states near Fermi level for HOMO and LUMO indicates they are formed by contributions from states of the same C atoms. Long et al. [163] and Liu et al. [159] also substantiated the metallic nature of this structure in their respective studies. However, unlike planar TG, the ambipolar 2D structure buckled TG shows Dirac-like Fermions attributed to π and π^* bands. This dissimilarity in the electronic properties can be described as follows. Planar TG consists of only one type of sublattice (one type of bonds) in the unit cell, whereas the buckled structure has two types of non-equivalent bonds. Nonetheless, Huang et al. [164] expressed their doubt on the existence of Dirac-like Fermions of buckled TG. They argued, however, that buckled TG is also normal metal and the band crossing near the Fermi level in buckled TG (as shown in Ref. [159]) can be derived from planar TG in terms of a different supercell by band folding (BF). Yet they pointed out that the carriers of buckled TG also show high Fermi velocity (v_F) of the order of $10^6 m/s$, which is a property of Dirac-like Fermions. Keeping all these in mind, a search of reversible hydrogen storage media leads to the investigation on Li-decorated planar TG [165]. This was modelled by one Li atom adsorbed on each (2×2) cell. For single adsorption, there are many possible choices like HS, BS and TS [53] as described in Figure 2.9. However, Li prefers to be adsorbed above the centre of the octagon (Figure 2.15), i.e. the HS of TG. Nearest Li–Li distance for planar TG (6.9 Å) is larger than that of graphene (4.92 Å) and Li doped B_2C sheet (5.12 Å). It is shown that

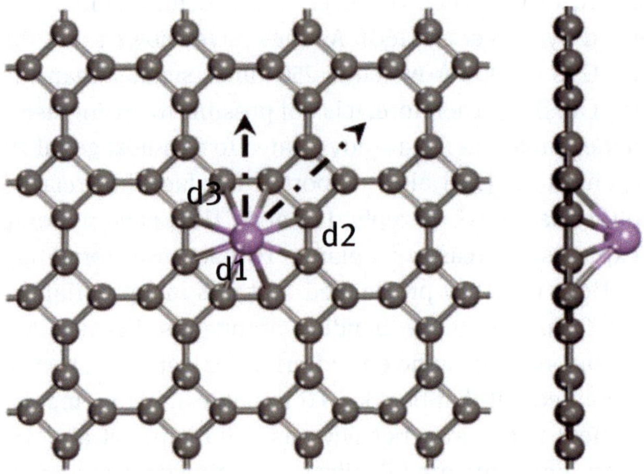

Figure 2.15: Favourable site for Li adsorption on planar TG. The arrows denote Li diffusion paths from the octagonal site to a neighbouring one.

each Li atom can adsorb hydrogen molecules. The underlying reason behind such adsorption can be described as follows. Li atoms get positively charged because of a charge transfer between Li and the sheet, which polarizes the H_2 molecule. So it is clear that, under such a process, TG can be functionalized into a feasible hydrogen storage medium. In addition, Liu et al. [161] explored that Li-decorated TG exhibits a high sensitivity to CO and the $Li-CO$ adsorption strength can be manipulated by external electrical field. Majidi [166] also has investigated the electronic properties of planar TG-like *CBN* and BN sheets and indicated these sheets are semiconductors in spite of the metallic nature of pristine planar TG.

2.5.2 TG nanoribbons (NRs) and clusters

Along with two types of TG sheets, Liu et al. [159] also investigated the variation of their properties with structural modifications, i.e. confinements. For buckled TG, armchair-like ribbon with width of one lattice constant (square-octagon periodic repetition along one direction) has the strongest quantum confinement and explores the linear dispersion relation. Whereas, periodic repetition of octagons, i.e. zigzag-like TGNR, exhibits metallic properties. Furthermore, spin-polarized first-principles calculations by them have indicated that zigzag-like TGNR prefers ferromagnetic state, while the former one is diamagnetic. Motivated by these intriguing properties of square-octagon repeated nets, Wang et al. [167] have employed first-principles study to investigate the structural and electronic properties of another tetra-symmetrical planar structure with space group P4/mmm. Lattice constant of the structure is

3.447 Å and its unit cell consists of 4 C atoms. Proper relaxation shows that, unlike graphene and the TG proposed by Liu et al., it consists of two distinct bonds (with bond lengths of 1.372 Å and 1.467 Å) and two different angles ($\frac{\pi}{2}$) and ($\frac{3\pi}{4}$). Absence of negative phonon modes and first-principles MD simulation at room temperature confirms its dynamical stability. Two types of charge density around two different bonds indicates a non-uniform charge distribution, which helps to designate these bonds as non sp^2 type. Band structure and finite DOS at Fermi level prefer this material as metallic and predict its sound potential for future nanoelectronics. They have also extended their work by calculating the width-dependent electronic properties of armchair and zigzag nanoribbons. They have mentioned the zigzag ribbon as uniformly metallic regardless of its width whereas armchair structure shows width-dependent odd–even metal–semiconductor oscillating behaviour. Those bandgaps also decrease with increasing width and are expected to vanish at infinite width (for sheet).

Later, Dai et al. [168] have investigated the transport properties of the TGNRs. In that work, they have also supported the width-dependent metal–semiconductor oscillations by calculating a general expression for the bandgap.

For further details about these tetrasymmetrical structures, first-principles based Raman and IR spectra have been investigated for different cluster sizes [169]. The study has revealed that a phonon Raman mode appearing near 1,711–1,713 cm^{-1} (shown in Figure 2.16) survives for different cluster sizes. The position (in cm^{-1}) agrees well with the previous reported phonon mode of planar TG sheet at 1,732 cm^{-1} by Wang et al. [167]. Therefore, this mode (shown in the inset(a) of Figure 2.16) must

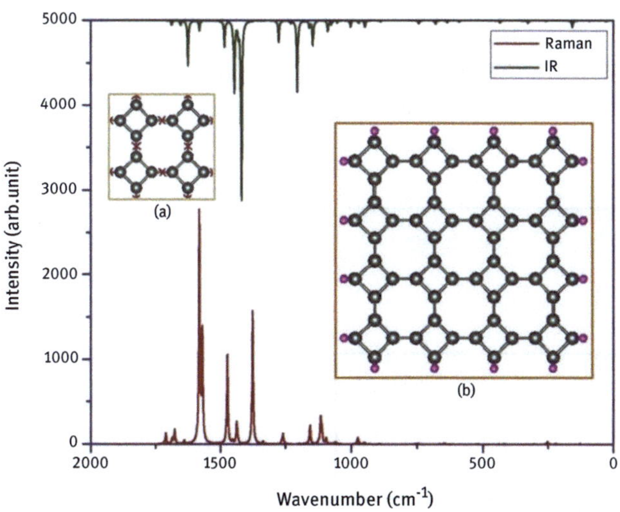

Figure 2.16: Raman and IR spectra of 4 × 4 TG cluster, characterizing mode of vibration for TG is shown in the inset (a), H passivated TG cluster is shown in inset (b).

be a characterizing Raman mode for pure planar TG. This method is also applied by Das et al. [170] to identify the G band of graphene in graphene quantum dots of variable sizes, which agrees well with the experimental descriptions. The characterizing mode for TG is a stretching mode which arises form the planar stretching of the bonds connecting the squares. Besides, the study of electronic properties has indicated that these pure TG clusters (dangling bonds are saturated by H atoms) do not show any dipole moment and possess zero DOS at E_F. Absence of DOS at E_F proves its semi-conducting nature. This gap decreases with increasing cluster size, which is well expected as for infinite size one must obtain the metallic property of planar TG. Furthermore, for a particular cluster size effect of B and N doping is also explored as B and N doping significantly tune the properties of graphene [171]. To start with, preferred doping cites (minimum energy position) for B and N are identified and are kept unchanged for rest of the study. Doping introduces asymmetry in charge distribution which results in finite dipole moment. B (N) doped structure shows relatively low (high) intense peak compared to pristine structure. Formation energy calculation indicates N doped structure is more stable than pristine and B doped structure. Vibrational details corresponding to prominent Raman modes are also reported in the study [169], for characterization purpose. Among those vibrational modes, low-intensity (low frequency as well) breathing like modes have been observed for both pristine and doped structures. IR spectra for these 3 × 3 TG clusters are shown in Figure 2.17. It is revealed that larger absorptions are caused by both in-plane and out-of-plane vibration. Corresponding wave numbers are 1,297.78 cm^{-1} (for in-plane vibration) and 48.90 cm^{-1} and 237.90 cm^{-1} (for out-of-plane vibration).

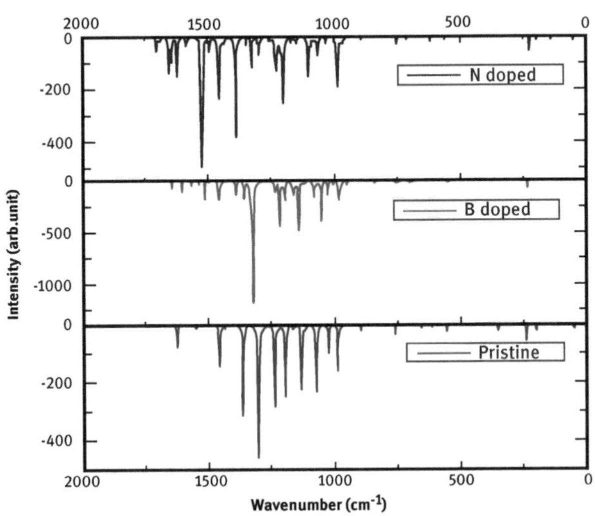

Figure 2.17: IR spectra of pristine and doped 3×3 TG clusters.

2.5.3 Other allotropes beyond TG

Apart from the above-mentioned TG allotrope, there are other competitors with overwhelming properties. Some of them are briefly discussed in this section.

Partial replacement of sp^2 hybridized aromatic bonds in graphene by carbyne chains forms a new family of graphene allotropes called graphynes. Baughman et al. [172] showed that different replacement portions of this acetylene linkage result in different types of graphynes. Among them, α, β and γ graphynes are well studied because of their high symmetric forms [173, 174]. Presence of Dirac cone in the energy band structure indicates semi-metallic behaviour of α and β graphynes. Whereas, γ graphyne is semiconducting in nature because of Kekule-distortion effect. Huang et al. [164] demonstrated the role of TB hopping parameters for graphynes in determining the condition for which Dirac cones can exist. Recently, successful syntheses of few graphynes are also reported [175].

Graphdiyne [176], first designed by Haley et al. [177], is composed of two acetylenic linkages between nearest-neighbour hexagonal rings and belongs to the space group P6mm. Later, many interesting properties like high degree of stability against temperature, semiconducting behaviour with silicon-like conductivity [178] and applicability in nanoscale devices [179] were reported. Interestingly, graphdiyne nanoribbons and nanotubes are experimentally synthesized and found to be useful in optoelectronic and spintronics [179, 180].

Another proposed allotrope is penta-graphene [181] which is composed of pentagons with C at its vertices, which closely resembles MacMohon's net, a semiregular tiling of the Euclidean plane similar to Cairo pentagonal tiling. Total thickness of this 2D multidecker sandwich sheet is found to be 1.2 Å. The structure possesses $P-42_1m$ symmetry and its tetragonal unit cell of lattice constant 3.64 Å consists of six atoms. Furthermore, the electronic properties confirm its indirect bandgap of 3.25 eV having semiconducting behaviour. Between rolled-up and staged penta-graphene, it is observed that penta-tubes are semiconducting regardless of its chirality. Despite all intriguing properties reported above, Ewels et al. [182] have strongly suggested that it will be impossible to construct penta-graphene experimentally. This allotrope cannot be isolated easily from some of its isomers because of their similar energies. Along with this, even a few catalytic impurities and environment effect however can force this structure to reconstruct rapidly towards graphene by continuous energy loss. In a recent study, Rajbanshi et al. have studied the properties of penta-graphene nanoribbons [183] and indicated that these are direct bandgap semiconductors with width-dependent tunable bandgap.

Stone–Wales (SW) defects are formed by a $\frac{\pi}{2}$ rotation of a carbon dimer with respect to the midpoint of the bond. Such SW defects in graphene lead towards the transformation of four hexagonal rings into pairs of pentagons and heptagons.

This transformed structure is also well studied and known as pentaheptites [184]. These structures are, in general, metallic in nature.

Another 2D allotropes can be achieved by systematic tessellations of octagons and pentagons (OP). Such alignments are of two types as predicted by Su et al. [185]. This can be viewed as colligating of ribbons formed by five-five-eight membered rings along a straight line path (for OPG-L) or along a zigzag path (OPG-Z). These structures are energetically more favourable than recently synthesized graphdiyne. Further studies on electronic properties indicate that OPG-L is a metal while OPG-Z is a gapless semi-metal.

Replacement of one-third sp^2 bonds of graphene by acetylenic linkages forms new set of energetically stable allotropes popularly known as Haeckelites as proposed by Terrones et al. [186]. This structure is a mixture of pentagons, hexagons or/and heptagons. There are three different members in this family: (a) symmetric arrangement of only heptagons and pentagons, (b) repeated structure of three connected heptagons, with alternating pentagons and hexagons, are in its surroundings and (c) tiling by connecting pentalene and heptalene and surrounded by hexagonal rings. In a later study, Enyashin et al. [39] have also introduced another periodic arrangement of pentagons, hexagons and octagons in this Haeckelites family and claimed its semiconducting behaviour.

Some other possible allotropes are also drawing keen attention of the researchers because of their novel exotic properties as follows. S-graphene is a periodic arrangement of six- and four-member rings with eight atoms in its unit cell. Similarly, D-graphene and E-graphenes are composed of $sp - sp^2$ and $sp^3 - sp^2$ hybridized C atoms, respectively [188]. In a review, Wang et al. [187] have explored that these tree 2D rectangular systems also exhibit Dirac cones. Another allotrope with tetra-rings and acetylenic linkages, known as rectangular graphyne (R-graphyne) [189], shows metallic behaviour. It is found that R-graphene nanoribbons can show both metallic, semiconducting and unexpected semi-mettalic properties depending on the shape and size of the ribbons. Two similarly proposed structures supergraphene and squarographenes are investigated to be metallic and show that relaxed squarographenes can be found in two forms, either arranged by distorted hexagons and regular squares [190] or by undistorted hexagons and rhombuses [186] while graphene-like material supergraphene is gapless semi-metal. A pictorial representation of the mentioned structures is given in Figure 2.18. Recently researchers, i.e. Kotakoski et al. and Lahiri et al., have used advanced experimental techniques like electron beam irradiation, defect engineering, etc. to form quasi-1D carbon structures consisting of repeated tetrarings-octarings and repeated octarings-pentarings [191, 192].

Although mostly electronic and structural properties in these materials are explored, however, studies involving the optical properties including Raman are still lacking. We hope to get a good direction if some of these materials are synthesized.

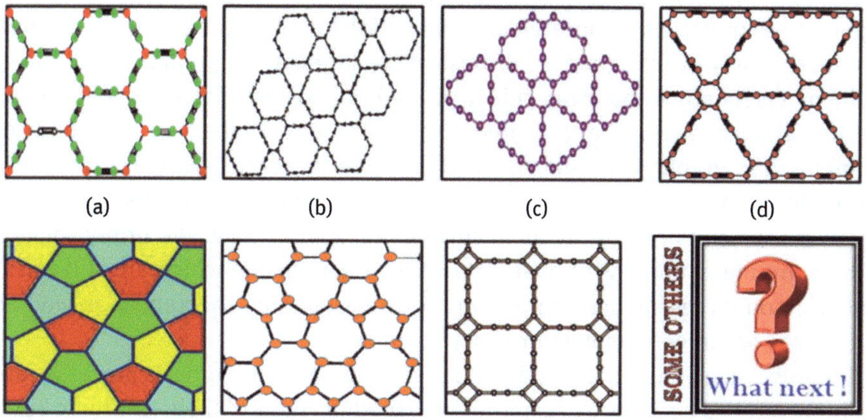

Figure 2.18: All graphene allotropes except TG (a) α graphyne, (b) β graphyne, (c) γ graphyne, (d) graphdiyne, (e) penta-graphene, (f) pentaheptites (g) R-graphene.

2.6 Conclusions and future directions

In this chapter, we have attempted to cover some important contributions of the electronic, magnetic and optical properties (including Raman spectroscopy) of silicene, germanene and T-graphene.

Silicene, one of the graphene's cousins, does not exist in nature in FS form; however, they can be synthesized on suitable metal single-crystal substrates. Tunability of the material properties in such elemental silicene sheets and disks (edge related) or nanoribbons offers a novel prospect of engineering discrete applications. Theoretical DFTB calculations have shown to indicate an appropriate direction to these silicene derivatives. Most of the silicene derivatives have been grown on metallic substrates; however, for device fabrication, these metallic substrates are not appropriate. It will be interesting to grow the various forms (single layer or multilayer) of silicene and its important derivatives on non-metallic or insulating substrates at a large scale. This will also aid to control the effective van der Waals interaction existing between the layers and the in-place lattice separation to modify effectively the band structure of Dirac fermions, if any, present in the structure. It is important to note that existing Si-based technologies recently is facing intrinsic limits with top-down approaches. It is therefore highly desirable that further experimental investigations should be pursued in order to exploit these theoretical predictions. These aspects can open in future a new horizon for low-dimensional, competitive and durable storage technologies.

Next to graphene and silicene, germanene stands as a fascinating material which can definitely be used in 2D nanoindustry in near future due to its unique and outstanding properties than graphene or silicene. Higher values of SOC and buckling designate germanene as a more efficient material for practical applications than graphene or silicene. Though it is not possible to fabricate FS germanene till date, germanene layers grown by using substrate material possess electronic and optical properties which can easily be compared with FS structure. Higher carrier mobility of germanene than graphene or silicene also manifest its practical applications. It is also possible to modify the properties of pristine germanene by means of different mechanisms which is one of the most current research interest. We firmly suggest that there are many unconventional properties of germanene which are still unexplored. This offers researchers a new way for investigation, theoretical as well as experimental, about germanene in nanoelectronics.

Although some preliminary studies both in theory and in synthesis processes are done, however, it is necessary to explore some modifications of their electronic band structure and optical properties in the presence of combinations of strains, doping and defects. Armed with theoretical predictions from DFT/TB, we strongly feel that there are unexpected properties of various structures made from silicene and germanene.

It is shown that structural modifications by changing the symmetry and tailoring the structure may influence the structural and electronic properties. For graphene allotropes, a class of materials show metallic behaviour, some of them are semiconducting and rest are semi-metals. This wide variety of bandgaps along with their appreciable structural stability directed towards the usefulness of these material in device formations. Among all the graphene allotropes, tetrasymmetrical graphene is dynamically more stable than graphyne and the recently prepared graphdiyne and has appreciable DOS at E_F. The bandgap of this material can be tuned by varying the ribbon width. In addition, TG sheet can be used as a gas storage media. Apart from these informations, other properties of these structures are relatively less explored. Hopefully, all the intriguing properties of TG and other graphene allotropes can be verified in real life by forming different optoelectronic devices, with the support of recent experimental progresses.

The unique intriguing features associated with these novel 2D structures beyond graphene are expected to overcome fundamental constraints imposed on dimensional physical limits on the evolution of faster, smaller and smart nanoelectronics devices in the near future.

Funding: This work is partially supported by DST-FIST, DST-PURSE, Government of India. Two of the authors (SC) and (ND) gratefully acknowledge DST, Government of India, for providing financial assistance through DST-INSPIRE Fellowship scheme of grant number IF120579 and IF150670.

References

[1] Novoselov KS, Mishchenko A, Carvalho A, Castro Neto AH. 2D materials and van der Waals heterostructures. Science. 2016;353:461–472.

[2] Novoselov KS, Geim AK, Morozov SV, Jiang D, Zhang Y, Dubonos SV, et al. Electric field effect in atomically thin carbon films. Science. 2004;306:666–669 .

[3] Novoselov KS, Geim AK, Morozov SV, et al. Two-dimensional gas of massless Dirac fermions in graphene. Nature. 2005;438:197–200.

[4] Geim AK, Grigorieva I. Van der Waals heterostructures. Nature. 2013;499:419–425.

[5] Gupta A, Sakthivel I, Seal S. Recent development in 2D materials beyond graphene. Prog Mater Sci. 2015;73:44–126.

[6] Jana D, Sun CL, Chen LC, Chen KH. Effect of chemical doping of boron and nitrogen on the electronic, optical, and electrochemical properties of carbon nanotubes. Prog Mater Sci. 2013;58:565–635.

[7] Balendhran S, Walia S, Nili H, Sriram S, Bhaskaran M. Elemental analogues of graphene: silicene, germanene, stanene, and phosphorene. Small. 2014;11:640–652.

[8] Kara A, Enriquez H, Seitsonen AP. A review on silicene-new candidate for electronics. Surf Sci Rep. 2012;67:1–18.

[9] Lew Yan, Voon LC, Guzmán-Verri GG. Is silicene the next graphene? MRS Bull., 2014;39:366–373.

[10] Xu M, Liang T, Shi M, Chen H. Graphene-like two-dimensional materials. Chem Rev. 2013;113:3766–3798.

[11] Yamada-Takamura Y, Friedlein R. Progress in the materials science of silicene. Sci Technol Adv Mater. 2014;15 064404 (12pp).

[12] Vogt P, Padova PD, Quaresima C, et al. Silicene: compelling experimental evidence for graphenelike two-dimensional silicon. Phys Rev Lett. 2012;108:155501–155504.

[13] Cinquanta E, Scalise E, Chiappe D, et al. Getting through the nature of silicene: an sp^2–sp^3 two-dimensional silicon nanosheet. J Phys Chem C. 2013;117:16719–16724.

[14] Lin CL, Arafune R, Kawahara K, et al. Structure of silicene grown on Ag(111). Appl Phys Exp. 2012;5 045802(3pp).

[15] Feng B, Ding Z, Meng S, et al. Evidence of silicene in honeycomb structures of silicon on Ag(111). Nano Lett. 2012;12:3507–3511.

[16] Chen L, Li H, Feng B, et al. Spontaneous symmetry breaking and dynamic phase transition in monolayer silicene. Phys Rev Lett. 2013;110 085504(5pp).

[17] Meng L, Wang Y, Zhang L, et al. Buckled silicene formation on Ir(111). Nano Lett. 2013;13:685–690.

[18] Fleurence A, Friedlein R, Ozaki T, Kawai H, Wang Y, Yamada-Takamura Y. Experimental evidence for epitaxial silicene on diboride thin films. Phys Rev Lett. 2012;108:245501–245504.

[19] Roome NJ, Carey JD. Beyond graphene: stable elemental monolayers of silicene and germanene. ACS Appl Mater Interfaces. 2014;6:7743–7750.

[20] Chowdhury S, Jana D. A Theoretical review on electronic, magnetic and optical properties of silicene. Rep Prog Phys. 2016;79 126501 (57pp).

[21] Zhuang J, Xu X, Feng H, Li Z, Wang X, Du Y. Honeycomb silicon: a review of silicene. Sci Bull. 2015;60:1551–1562.

[22] Oughaddou H, Enriquez H, Tchalala MR, et al. Silicene, a promising new 2D material. Prog Surf Sci. 2015;90:46–83.

[23] Ni Z, Liu Q, Tang K, et al. Tunable bandgap in silicene and germanene. Nano Lett. 2012;12:113–118.

[24] Drummond ND, Zólyomi V, Fal'ko VI. Electrically tunable band gap in silicene. Phys Rev B. 2012;85:075423–075427.

[25] Mohan B, Kumar A, Ahluwalia PK. Electronic, mechanical and dielectric properties of silicane under tensile strain. Phys E. 2013;53:233–239.
[26] Liu CC, Jiang H, Yao YG. Low-energy effective Hamiltonian involving spin-orbit coupling in silicene and two-dimensional germanium and tin. Phys Rev B. 2011;84:195430–195440.
[27] Liu CC, Feng W, Yao YG. Quantum spin hall effect in silicene and two-dimensional germanium. Phys Rev Lett. 2011;107 076802 (4pp).
[28] Qi XL, Zhang SC. Topological insulators and superconductors. Rev Mod Phys. 2011;83:1057–1110.
[29] Garcia JC, de Lima DB, Assali LVC, Justo JF. Group IV graphene- and graphane-like nanosheets. J Phys Chem C. 2011;115:13242–13246.
[30] Durgun E, Tongay S, Ciraci S. Silicon and III-V compound nanotubes: structural and electronic properties. Phys Rev B2005;72:075420 zcelik VO, Ciraci S. Self-assembly mechanisms of short atomic chains on single-layer graphene and boron nitride. Phys Rev B 2012;86:155421. Ataca C, Ciraci S. Functionalization of single-layer MoS2 honeycomb structures. J Phys Chem C2011;115:1330.
[31] Houssa M, van den Broek B, Scalise E, et al. Theoretical aspects of graphene-like group IV semiconductors. Appl Surf Sci. 2014;291:98–103.
[32] Qui J, Fu H, Xu Y, et al. From Silicene to Half-Silicane by Hydrogenation. ACS Nano. 2015;9:11192–11199.
[33] Wang W, Olovsson W, Uhrberg RIG. Band structure of hydrogenated silicene on Ag(111): evidence for half-silicane. Phys Rev B. 2016;93 081406(R)(4pp).
[34] Tao L, Cinquanta E, Chiappe D, et al. Silicene field-effect transistors operating at room temperature. Nature Nanotechnol. 2015;10:227–231.
[35] Tang Q, Zhou Z. Graphene-analogous low-dimensional materials. Prog Mater Sci. 2013;58:1244–1315.
[36] Liu H, Gao J, Zhao J. Silicene on substrates: a way to preserve or tune its electronic properties. J Phys Chem C. 2013;117:10353–10359.
[37] Kaloni TP, Schreckenbach G, Freund MS, Schwingenschlogl U. Current developments in silicene and germanene. Phys Status Solidi RRL. 2016;10:133–142.
[38] Butler SZ, Hollen SM, Cao L, et al. Progress, challenges, and opportunities in two-dimensional materials beyond graphene. ACS Nano. 2013;7:2898–2926.
[39] Enyashin AN, Ivanovskii AL. Graphene allotropes. Phys Status Solidi B. 2011;248:1879–1883.
[40] Heimann RB, Evsyukov SE, Koga Y. Carbon allotropes: a suggested classification scheme based on valence orbital hybridization. Carbon. 1997;35:1654–1658.
[41] Mas-Balleste R, Gomez-Navarro C, Gomez-Herrero J, Zamora F. 2D materials: to graphene and beyond. Nanoscale. 2011;3:20–30.
[42] Hohenberg P, Kohn W. Inhomogeneous electron gas. Phys Rev. 1964;136:B864–871.
[43] Kohn W, Sham LJ. Self-consistent equations including exchange and correlation effects. Phys Rev. 1965;140:A1133–1138.
[44] Kohn W. Nobel lecture: electronic structure of matter–wave functions and density functionals. Rev Mod Phys. 1999;71:1253–1266.
[45] Guistino F. Materials modelling using density functional theory. Oxford: Oxford University Press, 2014.
[46] Nath P, Chowdhury S, Sanyal D, Jana D. Ab-initio calculation of electronic and optical properties of nitrogen and boron doped graphene nanosheet. Carbon. 2014;73:275–282.
[47] Eberlein T, Bangert U, Nair RR, et al. Plasmon spectroscopy of free-standing graphene films. Phys Rev B. 2008;77 233406(4pp).
[48] Heyd J, Scuseria GE, Ernzerhof M. Hybrid functionals based on a screened Coulomb potential. J Chem Phys 2003;1188207 (9pp). Erratum: J Chem Phys. 2006;124(219906):1.

[49] Perdue JP, Burke K, Ernzerhof M. Generalized gradient approximation made simple. J Chem Phys. 1996;105 9982 (4 pp).
[50] Mahan GD. Many particle physics. New York, NY: Plenum Press, 1990.
[51] Dressel M, Gruner G. Electrodynamics of solids. Cambridge: Cambridge University Press, 2002.
[52] Nath P, Sanyal D, Jana D. Ab-initio calculation of optical properties of AA-stacked bilayer graphene with tunable layer separation. Curr Appl Phys. 2015;15:691–697.
[53] Nath P, Sanyal D, Jana D. Optical properties of transition metal atom adsorbed graphene: a density functional theoretical calculation. Phys E. 2015;69:306–315.
[54] Das R, Chowdhury S, Jana D. A first principles approach to magnetic and optical properties in single-layer graphene sandwiched between boron nitride monolayers. Mater Res Express. 2015;2 075601(12pp).
[55] Chowdhury S, Das R, Nath P, Jana D, Sanyal D. Electronic and optical properties of boron- and nitrogen-functionalized graphene nanosheet. In: Thakur VK, Thakur MK, editors. Chemical functionalization of carbon nanomaterials: chemistry and applications. New York, NY: CRC Press Chapter 42, 2015:949–957.
[56] Yu PY, Cardona M. Fundamentals of semiconductors physics and materials properties. New York, NY: Springer, 2010.
[57] Kamal C, Banerjee A, Chakrabarti A. In: Aliofkhazraei M, editors. Properties of two-dimensional silicon versus carbon systems graphene science handbook size-dependent properties. New York, NY: CRC Press. Chapter 15 221–234pp ISBN 9781466591356 – CAT# K20509 2016.
[58] Bao H, Guo J, Liao W, Zhao H. Electrical engineering of the optical properties in silicene. Appl Phys A. 2015;118:431–5.
[59] Hu W, Li Z, Structural Yang J. electronic, and optical properties of hybrid silicene and graphene nanocomposite. J Chem Phys. 2013;139 154704 (5pp).
[60] Mohan B, Kumar A, Ahluwalia PK. Electronic and optical properties of silicene under uni-axial and bi-axial mechanical strains: a first principle study. Phys E. 2014;61:40–47.
[61] Matthes L, Gori P, Pulci O, Bechstedt F. Universal infrared absorbance of two-dimensional honeycomb group-IV crystals. Phys Rev B. 2013;87 035438(9pp).
[62] Matthes L, Pulci O, Bechstedt F. Optical properties of two-dimensional honeycomb crystals graphene, silicene, germanene, and tinene from first principles. New J Phys. 2014;16 105007(13pp).
[63] Das R, Chowdhury S, Majumdar A, Jana D. Optical properties of P and Al doped silicene: a first principles study. RSC Adv. 2014;5:41–50.
[64] Ezawa M. Peculiar band gap structure of graphene nanoribbons. Phys Stat Sol C. 2007; 4:489–492.
[65] Rader HJ, Rouhanipour A, Talarico AM, Palermo V, Samor P, Mullen K. Processing of giant graphene molecules by soft-landing mass spectrometry. Nat Mater. 2006;5:276–280.
[66] Sony P, Shukla A. A general purpose Fortran 90 electronic structure program for conjugated systems using Pariser-Parr-Pople model. Comput Phys Commun. 2010;181:821–830.
[67] Chowdhury S, Nath P, Jana D. Shape dependent magnetic and optical properties in silicene nanodisks: a first principles study. J Phys Chem Sol. 2015;83:32–39.
[68] Majumdar A, Chowdhury S, Nath P, Jana D. Defect induced magnetism in planar silicene: a first principles study. RSC Adv. 2014;4:32221–32227.
[69] Wei W, Dai Y, Huang B, Jacob T. Many-body effects in silicene, silicane, germanene and germanane. Phys Chem Chem Phys. 2013;15 8789(6pp).
[70] Zakerian F, Berahman M. Optical properties of defected silicene: the many-body approach. Opt Quant Electron. 2016;48 370 (7pp).

[71] Ekardt W. Metal clusters. New York, NY: John Wiley and Sons Ltd., 1999.
[72] Yang JY, Liu LH. Temperature-dependent dielectric functions in atomically thin graphene, silicene, and arsenene. Appl Phys Lett. 2015;107 091902(4pp).
[73] Ye XS, Shao ZG, Zhao H, Yang L, Wang CL. Electronic and optical properties of silicene nanomeshes. RSC Adv. 2014;4:37998–38003.
[74] Guo L, Zhang S, Feng W, Hu G, Li W. A first-principles study on the structural, elastic, electronic, optical, lattice dynamical, and thermodynamic properties of zinc-blende CdX (X = S, Se, and Te). J Alloys Compd. 2013;579:583–593.
[75] Sugiyama Y, Okamoto H, Mitsuoka T, et al. Synthesis and optical properties of monolayer organosilicon nanosheets. J Am Chem Soc. 2010;132:5946–5947.
[76] Wang R, Xu M, Pi X. Chemical modification of silicene. Chin Phys B. 2015;24 086807 (21pp).
[77] Okamoto H, Nakano H, Sugiyama Y. Synthesis and modification of silicon nanosheets and other silicon nanomaterials. Chem Eur J. 2011;17:9864–9887.
[78] Nakano H, Nakano M, Nakanishi K, et al. Preparation of alkyl-modified silicon nanosheets by hydrosilylation of layered polysilane (Si_6H_6Si6H6). J Am Chem Soc. 2012;134: 5452–5455.
[79] Borensztein Y, Prevot G, Masson L. Large differences in the optical properties of a single layer of Si on Ag(110) compared to silicene. Phys Rev B. 2014;89 245410 (7pp).
[80] Matthes L, Pulci O, Bechstedt F. Massive dirac quasiparticles in the optical absorbance of graphene, silicene, germanene, and tinene. J Phys Condens Matter. 2013;25 395305(7pp).
[81] Bechstedt F, Matthes L, Gori P, Pulci O. Infrared absorbance of silicene and germanene. Appl Phys Lett. 2012;100 261906(4pp).
[82] Nair RR, Blake P, Griaorenko AN, et al. Fine structure constant defines visual transparency of graphene. Science. 2008;320 1308(9pp).
[83] Pulci O, Gori P, Marsili M, Garbuio V, Del Sole R, Bechstedt F. Strong excitons in novel two-dimensional crystals: silicane and germanane. Euro Phys Lett. 2012;98 37004(6pp).
[84] Kan E, Li Z, Yang J. Magnetism in graphene systems. NANO Brief Rep Rev. 2008;3:433–442.
[85] Zhang Z, Schwingenschlogl U, Rohan IS. Possible mechanism for d0 ferromagnetism mediated by intrinsic defects. RSC Adv. 2014;4:50759–50764.
[86] Coey JMD. d0 ferromagnetism. Solid State Sci. 2005;7:660–667.
[87] Hadacek N, Nossov A, Ranno L, Strobel P, Galra RM. Magnetic properties of HfO_2 thin films. J Phys Cond Matt. 2007;19 486206(12pp).
[88] Ghosh S, Khan GG, Das B, Mandal K. Vacancy-induced intrinsic d0 ferromagnetism and photoluminescence in potassium doped ZnO nanowires. J Appl Phys. 2011;109 123927(6pp).
[89] Peng C, Liang Y, Wang K, Zhang Y, Zhao G, Wang Y. Possible origin of ferromagnetism in an undoped ZnO d0 Semiconductor. J Phys Chem C. 2012;116:9709–9715.
[90] Ghosh S, Rakhishit T, Ranganathan R, Jana D. Role of Zn-interstitial defect states on d0 ferromagnetism of mechanically milled ZnO nanoparticles. RSC Adv. 2015;5:99766–99774.
[91] Palanichamy RR, Priyanga GS, Cinthia AJ, Murugan A, Meenaatci AT, Iyakutti K. Half metallic ferromagnetism in alkaline-earth metal nitrides XN (X=Ca, Sr and Ba): a first principles study. J Mag Mag Mat. 2013;346:26–37.
[92] Geshi M, Kusakabe K, Nagara H, Suzuki N. Synthetic ferromagnetic nitrides: First-principles calculations of CaN and SrN. Phys Rev B. 2007;76 054433 (5pp).
[93] Sahin H, Cahangirov S, Topsakal M, Bekaroglu E, Akturk E, R T Senger, Ciraci S. Monolayer honeycomb structures of group-IV elements and III-V binary compounds: first-principles calculations. Phys Rev B. 2009;80 155453(12pp).
[94] L B Drissi, E H Saidi, Bousmina M, Fassi-Fehri O. DFT investigations of the hydrogenation effect on silicene/graphene hybrids. J Phys Condens Matter. 2012;24 485502 (8pp).

[95] L B Drissi, Ramadana F Z. Many body effects study of electronic & optical properties of silicenegraphene hybrid. Phys E. 2015;68:38–41.
[96] Chowdhury S, Jana D. Electronic and magnetic properties of modified silicene/graphene hybrid: Ab initio study. Mat Chem Phys. 2016;183:580–587.
[97] Ju W, Li T, Su X, Cui H, Li H. Engineering magnetism and electronic properties of silicene by changing adsorption coverage. Appl Surf Sci. 2016;384:65–72.
[98] Zheng F, Zhang C. The electronic and magnetic properties of functionalized silicene: a first-principles study. Nanoscale Res Lett. 2012;7 422 (5pp).
[99] X Q Wang, H D Li, Wang J T. Induced ferromagnetism in one-side semihydrogenated silicene and germanene. Phys Chem Chem Phys. 2012;14:3031–3036.
[100] A P Paszkowska, Krawiec M. Electrical and mechanical controlling of the kinetic and magnetic properties of hydrogen atoms on free-standing silicene. J Phys Condens Matter. 2016;28 284004 (10pp).
[101] Zhang P, X D Li, C H Hu, S Q Wu, Zhu Z Z. First-principles studies of the hydrogenation effects in silicene sheets. Phys Lett A. 2012;376:1230–1233.
[102] Zhang C, Yan S. Scholarly articles for first-principles study of ferromagnetism in two-dimensional silicene with hydrogenation. J Phys Chem C. 2012;116:4163–4166.
[103] Ju W, Li T, Hou Z, Wang H, Cui H, Li X. Exotic d0 magnetism in partial hydrogenated silicene. Appl Phys Lett. 2016;108 212403 (5pp).
[104] Han N, Liu H, Zhao J. Novel magnetic monolayers of transition metal silicide. J Supercond Nov Magn. 2015;28:1755–1758.
[105] Le HM, Pham TT, Dinh TS, Kawazoe Y, Manh DN. First-principles modeling of 3d-transition-metal-atom adsorption on silicene: a linear-response DFT+ U approach. J Phys Condens Matter. 2016;28 135301 (11pp).
[106] Zheng R, Chen Y, Ni J. Highly tunable magnetism in silicene doped with Cr and Fe atoms under isotropic and uniaxial tensile strain. Appl Phys Lett. 2015;107 263104 (5pp).
[107] Sahin H, Peeters FM. Adsorption of alkali, alkaline-earth, and 3d transition metal atoms on silicene. Phys Rev B. 2013;87 085423 (9pp).
[108] Ozcelik VO, Ciraci S. Local reconstructions of silicene induced by adatoms. J Phys Chem C. 2013;117:26305–26315.
[109] Ozcelik VO, Gurel HH, Ciraci S. Self-healing of vacancy defects in single-layer graphene and silicene. Phys Rev B. 2013;88 045440 (11pp).
[110] Sun X, Wang L, Lin H, Hou T, Li Y. Induce magnetism into silicene by embedding transition-metal atoms. App Phys Lett. 2015;106 222401 (5pp).
[111] Ghosh D, Parida P, Pati SK. Stable line defects in silicene. Phys Rev B. 2015;92 195136 (11pp).
[112] Zhao T, Zhang S, Wang Q, Kawazoe Y, Jena P. Tuning electronic and magnetic properties of silicene with magnetic superhalogens. Phys Chem Chem Phys. 2014;16:22979–22986.
[113] Ezawa M. Metallic graphene nanodisks: Electronic and magnetic properties. Phys Rev B. 2007;76 245415 (6pp).
[114] Ezawa M. Graphene nanoribbon and graphene nanodisk. Phys E. 2008;40:1421–1423.
[115] Kikutake K, Ezawa M, Nagaosa N. Edge states in silicene nanodisks. Phys Rev B. 2013;88 115432 (6pp).
[116] Martinazzo R In: Aliofkhazraei M, Ali N, Milne WI, et al., editors. Atomic-scale defects and impurities in graphene. New York, NY:: CRC Press, Taylor & Francis, 2016:21–37. Graphene science handbook nanostructure and atomic arrangement.
[117] C C Liu, Jiang H, Yao Y. Low-energy effective Hamiltonian involving spin-orbit coupling in silicene and two-dimensional germanium and tin. Phys Rev B. 2011;84 195430(11pp).
[118] Cahangirov S, Topsakal M, Akturk E, Sahin H, Ciraci S. Two- and one- dimensional honeycomb structures of silicon and germanium. Phys Rev Lett. 2009;102 236804 (4pp).

[119] Nijamudheen A, Bhattacharjee R, Choudhury S, Datta A. Electronic and chemical properties of germanene: the crucial role of buckling. J Phys Chem C. 2015;119:3802–3809.

[120] Kaloni T P. Tuning the structural, electronic, and magnetic properties of germanene by the adsorption of 3d transition metal atoms. J Phys Chem C. 2014;118:25200–25208.

[121] Sun M, Ren Q, Wang S, Zhang Y, Du Y, Yu J, Tang W. Magnetism in transition-metal-doped germanene: a first-principles study. Comput Mater Sci. 2016;118:112–116.

[122] Deng Z, Li Z, Wang W. Electron affinity and ionization potential of two- dimensional honeycomb sheets: a first principle study. Chem Phys Lett. 2015;637:26–31.

[123] J E Padilha, Pontes R B. Electronic and transport properties of struc- tural defects in monolayer germanene: an ab initio investigation. Solid State Commun. 2016;225:38–43.

[124] X –S Ye, Z –G Shao, Zhao H, Yang L, Wang C –L. Intrinsic carrier mobility of germanene is larger than graphenes: first-principle calculations. RSC Adv. 2014;4:21216–21220.

[125] Scalise E, Houssa M, Pourtois G. Broek B v d, Afanasev V, Stesmans A. Vibrational properties of silicene and germanene. Nano Res. 2013;6:19–28.

[126] Davila ME, Xian L, Cahangirov S, Rubio A, Lay GL. Germanene: a novel two-dimensional germanium allotrope akin to graphene and silicene. New J Phys. 2014;16 095002 (11pp).

[127] Li L, Lu S-Z, Pan J, Qin Z, Wang Y-Q, Wang Y, Cao G-Y, Du S, Gao H-J. Buckled germanene formation on Pt(111). Adv Mater. 2014;26:4820–4824.

[128] Liu G, Liu SB, Xu B, Ouyang CY, Song HY. First-principles study of the stability of free-standing germanene in oxygen atmosphere. J Appl Phys. 2015;118 124303 (7pp).

[129] Davila ME, Lay GL. Few layer epitaxial germanene: a novel two-dimensional Dirac material. Sci Rep. 2016;6 20714 (9pp).

[130] Miro P, Audiffred M, Heine T. An atlas of two-dimensional materials. Chem Soc Rev. 2014;43:6537–6554.

[131] Asano K, Hotta C. Designing Dirac points in two-dimensional lattices. Phys Rev B. 2011; 83 245125 (14pp).

[132] Acun A, Zhang L, Bampoulis P, et al. Germanene: the germanium analogue of graphene. J Phys Condens Matter. 2015;27 443002 (12pp).

[133] Zhang L, Bampoulis P, Houselt AV, Zandvliet HJ. Two-dimensional Dirac signature of germanene. Appl Phys Lett. 2015;107 111605 (5pp).

[134] Behera H, Mukhopadhyay G. First principles study of structural and elec- tronic properties of germanene. AIP Conf Proc. 2011;1349:823–824.

[135] Houssa M, Pourtois G, Afanasev VV, Stesmans A. Electronic properties of two-dimensional hexagonal germanium. Appl Phys Lett. 2010;96 082111 (4pp).

[136] Lebegue S, Eriksson O. Electronic structure of two-dimensional crystals from ab initio theory. Phys Rev B. 2009;79 115409 (4pp).

[137] Walhout CJ, Acun A, Zhang L, Ezawa M, Zandvliet HJ. Scanning tunneling spectroscopy study of the Dirac spectrum of germanene. J Phys Condens Matter. 2016;28 284006 (7pp).

[138] Li P, Cao J, Guo Z-X. New approach for fabrication germanene with Dirac electrons preserved: a first principle study. J Mater Chem C. 2016;4:1736–1740.

[139] Houssa M, Scalise E, Sankaran K, Pourtois G, Afanasev VV, Stesmans A. Electronic properties of hydrogenated silicene and germanene. Appl Phys Lett. 2011;98 223107 (4pp).

[140] Rupp CJ, Chakraborty S, Ahuja R, Baierle RJ. The effect of impurities in ultra-thin hydrogenated silicene and germanene: a first principles study. Phys Chem Chem Phys. 2015;17:22210–22216.

[141] Zberecki . Emergence of magnetism in doped two-dimensional honeycomb structures of IIIV binary compounds. J Supercond Nov Magn. 2012;25:2533–2577.

[142] Li S-J, Su Y, Chen G. Patterning germanene into superlattices: an efficient method for tuning conducting properties. Chem Phys Lett. 2015;638:187–190.

[143] Liang P, Liu Y, Xing S, Shu H, Tai B. Electronic and magnetic prop- erties of germanene: Surface functionalization and strain effects. Solid State Commun. 2016;226:19–24.

[144] Pang Q, Zhang C-L, Li L, Fu Z-Q, Wei X-M, Song Y-L. Adsorption of alkali metal atoms on germanene: a first-principles study. Appl Surf Sci. 2014;314:15–20.

[145] Pang Q, Li L, Zhang C-L, Wei X-M, Song Y-L. Structural, electronic and magnetic properties of 3d transition metal atom adsorbed germanene: a first-principles study. Mater Chem Phys. 2015;160:96–104.

[146] Zaveh SJ, Roknabadi MR, Morshedloo T, Modarresi M. Electronic and thermal properties of germanene and stanene by first-principles calculations. Superlattices Microstruct. 2016;91:383–390.

[147] Li SS, Zhang CW, Ji WX, Li F, Wang PJ, Hu SJ, S S Yan, Liu Y S. Tunable electronic and magnetic properties in germanene by alkali, alkaline-earth, group III and 3d transition metal atom adsorption. Phys Chem Chem Phys. 2014;16:15968–15978.

[148] Xia W, Hu W, Li Z, Yang J. A first-principles study of gas adsorption on germanene. Phys Chem Chem Phys. 2014;16:22495–22498.

[149] Ozcelik VO, Durgun E, Ciraci S. New phases of germanene. J Phys Chem Lett. 2014;5:2694–2699.

[150] Gurel HH, Ozcelik VO, Ciraci S. Effects of charging and perpendicular electric field on the properties of silicene and germanene. J Phys Condens Matter. 2013;25 305007 (7pp).

[151] Ye M, Quhe R, Zheng J, Ni Z, Wang Y, Yuan Y, Tse G, Shi J, Gao Z, Lu J. Tunable band gap in germanene by surface adsorption. Phys E. 2014;59:60–75.

[152] Dhar N, Bandyopadhyay A, Jana D. Tuning electronic, magnetic and optical properties of germanene nanosheet with site dependent adatoms arsenic and gallium: a first principles study. Curr Appl Phys. 2017;17:573–583.

[153] Tomoyose T. Dielectric constants of Ge, Sn and isoelectronic semiconductors. J Phys Soc Jap. 1994;63:1149–1155.

[154] Icenogle HW, Platt Ben C, Wolfe WL. Refractive indexes and temperature coefficients of germanium and silicon. Appl Opt. 1976;15:2348–2351.

[155] Wei W, Dai Y, Huang B, Jacob T. Many-body effects in silicene, silicane, germanene and germanane. Phys Chem Chem Phys. 2013;15:8789–8794.

[156] Rast L. In: Misra P, editors. Plasmonic properties of metallic nanostructures, two dimensional materials, and their composites. Singapore: Springer science + Business media, 2015: 165–189. Applied spectroscopy and the science of nanomaterials, progress in optical science and photonics.

[157] Rast L, Tewary VK. Electron energy loss function of silicene and germanene multilayers on silver., 2013;arXiv:1311.0838v.

[158] Zhu H, Balaban AT, Klein DJ, Zivkovic TP. Conjugated circuit computations on two dimensional carbon networks. J Chem Phys A. 1994;101:5281–5292.

[159] Liu Y, Wang G, Huang Q, Guo L, Chen X. Electron-phonon couplings and carrier mobility in graphynes. Phys Rev Lett. 2012;108 225505(5).

[160] Kim BG, Jo JY, Sim HS. Comment on structural and electronic properties of T graphene: a two-dimensional carbon allotrope with tetrarings. Phys Rev Lett. 2013;110 029601(1pp).

[161] Liu CS, Jia R, Ye XJ, Zeng ZJ. Non-hexagonal symmetry-induced functional T graphene for the detection of carbon monoxide. J Chem Phys. 2013;139 034704(6pp).

[162] Liu Y, Wang G, Huang Q, Guo L, Chen X. Liu et al. Reply. Phys Rev Lett. 2013;110 029602(1).

[163] Long G, Zhou Y, Jin M, et al. Theoretical investigation on two-dimensional non-traditional carbon materials employing three-membered ring and four-membered ring as building blocks. Carbon. 2015;95:1033–1038.

[164] H Q Huang, Y C Li, Z R Liu, Wu J, Duan W H. Preceding comment. Phys Rev Lett. 2012;109 269601(1pp).

[165] Ye XJ, Liu CS, Zhong W, Zeng Z, Du YW. Metalized T graphene: A reversible hydrogen storage material at room temperature. J Appl Phys. 2014;116 114304(4pp).
[166] Majidi R. Electronic properties of T graphene-like CBN sheets: a density functional theory study. Phys E. 2015;74:371–376.
[167] Wang XQ, Li HD, Wang JT. Structural stabilities and electronic properties of planar C_4C4 carbon sheet and nanoribbons. Phys Chem Chem Phys. 2012;14:11107–11111.
[168] Dai CJ, Yan XH, Xiao Y, Guo YD. Electronic and transport properties of T-graphene nanoribbon: Symmetry-dependent multiple dirac points, negative differential resistance and linear current-bias characteristics. Euro Phys Lett. 2014;107 37004(6pp).
[169] Bandyopadhyay A, Pal P, Chowdhury S, Jana D. First principles Raman study of boron and nitrogen doped planar T-graphene clusters. Mater Res Express. 2015;2 095603(12).
[170] Das R, Dhar N, Bandyopadhyay A, Jana D. Size dependent magnetic and optical properties in diamond shaped graphene quantum dots: A DFT study. J Phys Chem Solids. 2016;99: 34–42.
[171] Jana D, Nath P, Sanyal D In: Aliofkhazraei M, Ali N, Milne WI, et al., editors. Modification of electronic properties of graphene by boron (B) and nitrogen (N) substitution. New York NY:: CRC Press, Taylor & Francis, 2016:231–246. Graphene science handbook nanostructure and atomic arrangement.
[172] Baughman RH, Eckhardt H, Kertesz M. Structure property predictions for new planar forms of carbon: layered phases containing sp^2 and sp atoms. J Chem Phys. 1987;87: 6687–6699.
[173] Kim BG, Choi HJ. Graphyne: Hexagonal network of carbon with versatile Dirac cones. Phys Rev B. 2012;86 115435(5pp).
[174] Huang HQ, Duan WH, Liu ZR. The existence/absence of Dirac cones in graphynes. New J Phys. 2013;15 023004(13pp).
[175] Li Y, Xu L, Liu H, Li Y. Graphdiyne and graphyne: from theoretical predictions to practical construction. Chem Soc Rev. 2014;43:2572–2586.
[176] Haley MM, Brand SC, Pak JJ. Carbon networks based on dehydrobenzoannulenes: synthesis of graphdiyne substructure. Angew Chem Int Ed. 1997;36:836–838.
[177] Haley MM, Pak JJ, Brand SC. Macrocyclic Oligo(phenylacetylenes) and Oligo(phenyldiacetylenes). Top Curr Chem. 1999;201:81–130 .
[178] Shao ZG, Ye XS, Yang L, Wang CL. The optical and electrical properties of silver nanowire mesh films. J Appl Phys. 2013;114 093712(3pp).
[179] Jalili S, Houshmand F, Schofield J. Study of carrier mobility of tubular and planar graphdiyne. J Appl Phys A Mater Sci Process. 2015;119:571–579.
[180] Li G, Li Y, Qian X, Liu H, Lin H, Chen N, Li Y. Construction of tubular molecule aggregations of graphdiyne for highly efficient field emission. J Phys Chem C. 2011;115:2611–2615.
[181] Zhang S, Zhou J, Wang Q, Chen X, Hawazoe Y, Penta-graphene: Jena P. a new carbon allotrope. Proc Nat Acad Sci. 2015;112:2372–2377.
[182] Ewels CP, Rocquefelte X, Kroto HW, Rayson MJ, Briddon PR, Heggie MI. Predicting experimentally stable allotropes: instability of penta-graphene. Proc Nat Acad Sci. 2015;112: 15609–15612.
[183] Rajbanshi B, Sarkar S, Mandal B, Sarkar P. Energetic and electronic structure of penta-graphene nanoribbons. Carbon. 2016;100:118–125.
[184] Crespi VH, Cohen ML. Prediction of a pure-carbon planar covalent metal. Phys Rev B. 1996;53: R13303–13305.
[185] Su C, Jiang H. Feng. Two-dimensional carbon allotrope with strong electronic anisotropy. J Phys Rev B. 2013;87 075453(5pp).

[186] Terrones H, Terrones M, Hernndez E, Grobert N, J C Charlier, Ajayan P M. New metallic allotropes of planar and tubular carbon. Phys Rev Lett. 2000;84:1716–1719.

[187] Wang J, Deng S, Liu Z, Liu Z. The rare two-dimensional materials with Dirac cones. Natl Sci Rev. 2015;2:22–39.

[188] Xu LC, Wang RZ, Miao MS, et al. Two dimensional Dirac carbon allotropes from graphene. Nanoscale. 2014;6:1113–1118.

[189] Yin WJ, Xie YE, Liu LM, Wang RZ, Wei XL, Lau L, Zhonga JX, Chena YP. R-graphyne: a new two-dimensional carbon allotrope with versatile Dirac-like point in nanoribbons. J Mater Chem A. 2013;1:5341–5346.

[190] Deza M, Fowler PW, Shtogrin M, Vietze K. Pentaheptite modifications of the graphite sheet. J Chem Inf Comput Sci. 2000;40:1325–1332.

[191] Kotakoski J, Krasheninnikov AV, Kaiser U, Meyer JC. From point defects in graphene to two-dimensional amorphous carbon. Phys Rev Lett. 2011;106 105505(4pp).

[192] Lahiri J, Lin Y, Bozkurt P, Oleynik II, Batzill M. An extended defect in graphene as a metallic wire. Nat Nanotech. 2010;5:326–329.

Toufik Khairat, Mohammed Salah, Khadija Marakchi
and Najia Komiha

3 Theoretical study of the electronic states of newly detected dications. Case of MgS^{2+} AND SiN^{2+}

Abstract: The dications MgS^{2+} and SiN^{2+}, experimentally observed by mass spectroscopy, are theoretically studied here. The potential energy curves of the electronic states of the two dications MgS^{2+} and SiN^{2+} are mapped and their spectroscopic parameters determined by analysis of the electronic, vibrational and rotational wave functions obtained by using complete active space self-consistent field (CASSCF) calculations, followed by the internally contracted multi-reference configuration interaction (MRCI)+Q associated with the AV5Z correlation consistent atomic orbitals basis sets. In the following, besides the characterization of the potential energy curves, excitation and dissociation energies, spectroscopic constants and a double-ionization spectra of MgS and SiN are determined using the transition moments values and Franck–Condon factors. The electronic ground states of the two dications appear to be of $X^3\Sigma^-$ nature for MgS^{2+} and $X^4\Sigma^-$ for SiN^{2+} and shows potential wells of about 1.20 eV and 1.40 eV, respectively. Several excited states of these doubly charged molecules also depicted here are slightly bound. The adiabatic double-ionization energies were deduced, at 21.4 eV and 18.4 eV, respectively, from the potential energy curves of the electronic ground states of the neutral and charged species. The neutral molecules, since involved, are also investigated here. From all these results, the experimental lines of the mass spectra of MgS and SiN could be partly assigned.

Keywords: electronic structure calculations, ab initio methods, potential energy curves, electronic excited states, ro-vibrational spectra, spectroscopic constants

3.1 Introduction

In order to participate in forming a spectroscopic database of the species potentially existing in the interstellar medium, we look to determine the electronic structures, the potential energy curves (PECs) and the spectroscopic constants of species recently detected. In this context, we have studied here the MgS^{2+} and SiN^{2+} dications detected in laboratory by Franzreb and Williams (Arizona State University) (see Figure 3.1) [1]. These species are shown to be long lived and metastable in the gas phase [2]. The theoretical study performed here is based on ab initio quantum chemistry calculations. The ab initio calculations, the theoretical determination of electronic structures, the potential

https://doi.org/10.1515/9783110467215-003

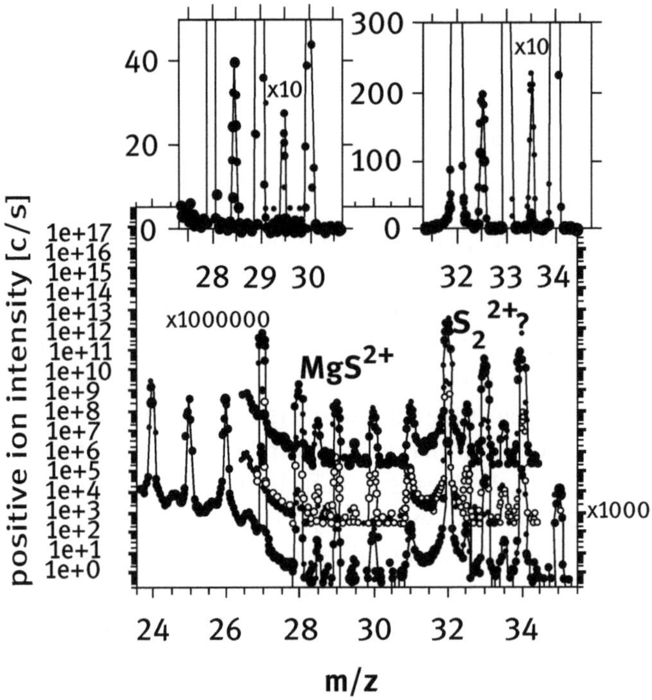

Figure 3.1: Mass spectrum of MgS^{2+} detected at half integer m/z 28. With the permission of authors of Ref [1].

energy curves and spectroscopic constants are definitely helpful to confirm the experimental results obtained by mass spectroscopy. We first investigate the neutral MgS and SiN molecules, and then we looked at the dications. The accurate ab initio calculations performed on the low-lying electronic states show the existence of, at least, 12 bound states of MgS and five bound states for MgS^{2+}, six electronic states for the neutral SiN and six for its dication. The potential energy curves of all these states are mapped, and the spectroscopic constants, the transition moments and the double photoionization spectra are determined. A comparison with experiment is made and an investigation on the spectroscopy and dissociation dynamics as well. Some spectroscopic constants and vibrational levels of the neutral electronic states (**Tables 3.1 and 3.2**) are also calculated here.

3.2 Computational details

Ab initio calculations of the potential energy curves of the different electronic states are performed using the CASSCF method (complete active space

Table 3.1: MRCI spectroscopic constants of electronic states of MgS, including the harmonic wavenumber (ω_e), the anharmonic terms (ω_{exe}, ω_{eye}), the rotational constants (β_e, α_e, γ). All values are in cm^{-1}. Equilibrium distances (R_e, in bohr), T_e (in eV) correspond to the MRCI adiabatic excitation energy, D_e (in eV) is calculated as the difference between the energy of the electronic state at equilibrium and the energy at the maximum of the potential barrier.

State	ω_e	ω_{exe}	ω_{eye}	β_e	α_e*	γ	R_e	T_e	D_e
$X^1\Sigma^+$	512.61	−2.27	−0.05	0.26	1d-03	-6d-05	4.103	0.00	3.02
	512.00**						4.090**		2.27**
	528.74***						4.049***		≤2.4***
	—						4.055****		2.86 ± 0.69*****
	511.00a						4.100a		2.30a
	524.00b								4.064b
$1^1\Pi$	431.25	−1.44	−0.04	0.22	1d-03	00.00	4.434	0.57	2.50
	431.00a						4.380a		1.70a
	422.00**						4.450**		1.86**
	449.00b						4.386b		
$2^1\Pi$	181.45	0.21	−0.17	0.09	7d-05	4d-05	7.135	4.86	1.06
	181.00**						6.920**		0.88**
$2^1\Sigma^+$	488.87	−3.28	0.05	0.25	1d-03	-3d-05	4.199	2.75	1.90
	489.00**						4.190**		1.99**
	497.34***						4.149***		
$1^1\Delta$	384.15	−5.09	−2.78	0.21	−0.015	-6d-03	4.601	3.45	—
	384.00**						4.450**		
$2^1\Delta$	141.31	−1.90	0.95	0.08	-7d-03	-1d-03	7.59	3.54	0.27
$3^1\Delta$	500.19	−17.38	0.22	0.17	4d-04	-2d-04	5.109	4.93	1.09
$1^3\Pi$	426.06	−2.93	0.001	0.22	1d-03	1d-05	4.412	0.30	1.63
	426.00**						4.400**		1.82**
	415.00c						4.420a		

(continued)

Table 3.1: (*continued*)

State	ω_e	ω_{exe}	ω_{eye}	β_e	α_e*	γ	R_e	T_e	D_e
	427.00d						4.391b		
$2^3\Pi$	232.02	−22.22	0.54	0.12	3d-03	1d-04	5.903	3.21	1.63
	232.00**							6.150**	1.60**
$1^3\Sigma^+$	413.66	−11.07	1.04	0.24	5d-03	5d-04	4.290	1.39	3.26
	413.00**						4.350**		3.25**
$2^3\Delta$	486.76	−19.59	1.83	0.23	8d-03	1d-03	4.314	3.29	1.36
$2^3\Sigma^-$	453.39	−3.78	−0.15	0.21	2d-03	-1d-04	4.544	3.84	1.01
	454.00**						4.530**		1.13**
$2^3\Pi$	232.02	−22.22	0.54	0.12	3d-03	1d-04	5.903	3.21	1.63
	232.00**							6.150**	1.60**
$1^3\Sigma^+$	413.66	−11.07	1.04	0.24	5d-03	5d-04	4.290	1.39	3.26
	413.00**						4.350**		3.25**
$2^3\Delta$	486.76	−19.59	1.83	0.23	8d-03	1d-03	4.314	3.29	1.36
$2^3\Sigma^-$	453.39	−3.78	−0.15	0.21	2d-03	-1d-04	4.544	3.84	1.01
	454.00**						4.530**		1.13**

* Vertical energy difference at 4.2 bohr.
** Theor., Ref [9].,
*** Exp., Ref [10].,
**** Exp., Ref [11].,
***** Exp., Ref [13].,
[a]Theor., Ref [12].,
[b]Theor., Ref [14].

Table 3.2: Vibrational levels of the different electronic states of MgS. All values are in cm^{-1}.

v	$X^1\Sigma^+$	$1^1\Pi$	$2^1\Pi$	$2^1\Sigma^+$	1^1	2^1	3^1	$1^3\Pi$	$2^3\Pi$	$1^3\Sigma^+$	2^3	$2^3\Sigma^-$
0	255.62	215.38	90.86	243.37	192.23	202.92	242.42	212.13	103.42	205.27	249.7	225.61
1	763.51	643.64	272.2	725.83	557.18	343.52	708.57	632.33	292.74	600.18	703.24	670.94
2	1,266.36	1,068.68	452.48	1,202.15	886.93	488.86	1,141.94	1,046.7	442.47	982.29	1,134.1	107.36
3	1,763.85	1,490.29	630.71	1,672.64	164.8	644.64	1,543.85	455.15	555.83	1,357.9	1,553.3	1,534
4	2,255.95	1,908.09	805.9	2,137.17	1,553.4	810.09	1,917.91	1,885.2	666.66	1,732	1,966.2	1,949.1
5	2,742.72	2,321.61	1,977.54	2,595.64	1,996.26	984.19	2,273.4	2,256.2	7,799.4	2,108.9	2,373.1	2,351.2
6	3,224.17	2,730.41	1,145.61	3,047.83	2,467.44	1,165.97	2,619.45	2,649.5	947.66	2,488.6	2,773.1	2,740.7
7	3,700.27	3,134.26	1,310.23	3,493.52	2,970.44	1,354.11	2,960.43	3,038.5	1,101.1	2,869.3	3,165.9	3,121.8
8	4,171.06	3,533.26	1,471.76	3,932.42	3,505.77	1,545.58	3,292.25	3,423.6	1,255.5	3,250.2	3,551.3	3,500.3
9	4,636.69	3,927.65	1,630.72	4,364.22	4,069.19	1,732.95	3,607.74	3,804.8	1,413.1	3,630.9	3,929.2	3,877.3
10	5,097.43	4,317.65	1,787.67	4,788.52	4,657.55	1,881.74	3,897.59	4,182.4	1,587.4	4,010.1	4,300.4	4,249.3

self-consistent field) in which all possible electronic excitations resulting for the distributions of the valence electrons into the valence molecular orbitals (MOs) were allowed. These valence MOs, calculated in the C_{2v} symmetry group, were correlated and all the electronic states of the same spin multiplicity were averaged together in the CASSCF calculations. These first CASSCF calculations are for determining the most important electronic states of the neutrals and the dications. The extended AV5Z basis sets used for Mg, S, Si and N atoms include diffuse orbitals in order to describe properly the upper electronic states supposed to be of Rydberg character. The augmented cc-pV5Z basis set constituted of orbitals of s,p,d,f,g,h, type could be considered as complete. These calculations were then followed by the MRCI ones (internally contracted multi-reference configuration interaction) [3] with AV5Z quality of atomic orbital basis sets as implemented in the Molpro package [4]. The inner MOs were frozen in these MRCI calculations, and all the valence MOs are included. The electronic states correlating to lowest asymptotes of MgS and SiN and their respective dications are determined. For MRCI calculations, the references are all configurations of the CI expansion of the CASSCF wave functions. In these reference spaces, the total number of configuration functions is 328 meaning 240 configurations for MgS and for SiN 254 contracted Gaussian type orbitals (cGTOs) are to be taken into account; this implies more than one million of contracted configurations in the whole CI calculations. The nuclear motion problem was then solved using the derivatives of the potential curves near the minimum energy distances. Standard perturbation calculations of the Cooley method were performed [5] and variational treatments [6] as well. The spectroscopic data of the bound states were then deduced. The potential energy curves of the lowest electronic states of the two neutral molecules and their dications have been used to calculate the Franck–Condon (FC) factors and simulate the double-photoionization spectra of MgS and SiN. The Level program [7] allows such a simulation. The methods used here are supposed to be accurate enough for the description of the relative energies of the computed electronic states.

3.3 Results and discussion

3.3.1 Neutral MgS

Figure 3.2 displays the global potential energy curves of low-lying singlet and triplet electronic states of the MgS molecule. These electronic states are correlated to the four lowest dissociation asymptotes: Mg (^3P_u) + S (^3P_g), Mg (^1S_g) + S(^1D_g), Mg (^3P_u) + S (^3P_g) and Mg (^1S_g) + S (^3P_g). The energy ordering of the asymptotes is deduced from the literature [8] and compared to the values from our calculations. As shown by V.W. Ribas et al. [9] (Table 3.3), the ground state is of $X^1\Sigma^+$ symmetry and presents a multiconfigurational character with an important contribution of the two

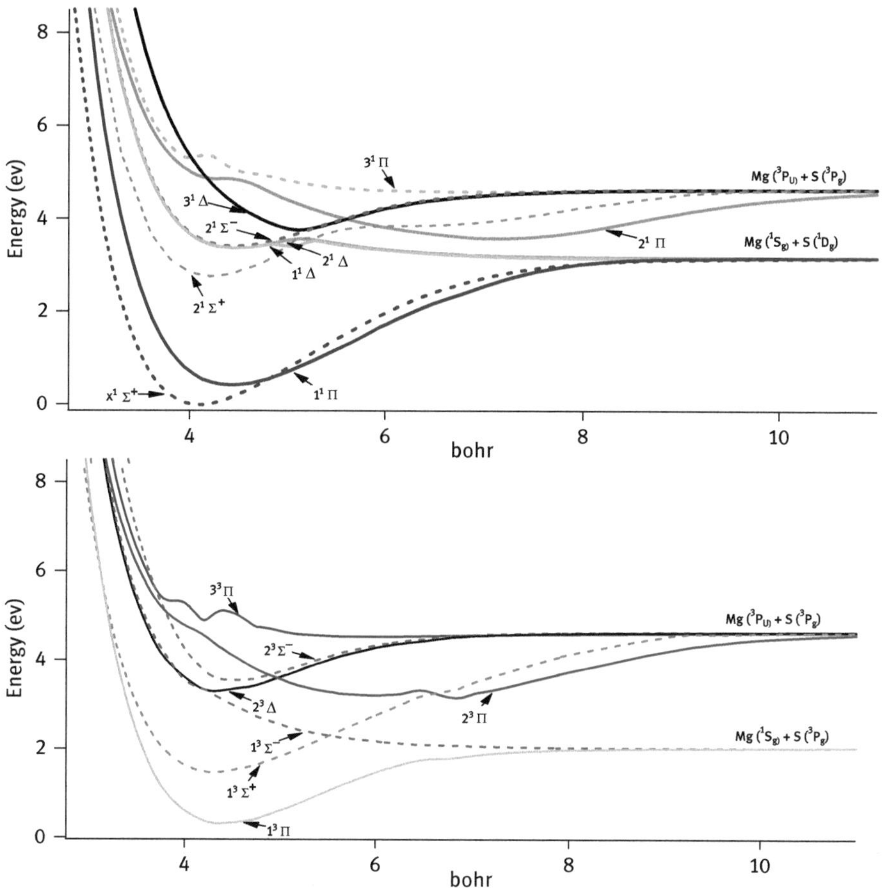

Figure 3.2: The potential curves of the neutral MgS. Singlet (lower panel) and triplet states (upper panel) are separately represented.

Table 3.3: Energies of the three lowest dissociation limits of MgS.

Asymptotes	Values from Nist (eV)	MRCI/AVZ (eV)
Mg(^{1}Sg) + S(^{3}Pg)	0.00	0.00
Mg(^{1}Sg) + S(^{1}Dg)	1.15	1.36
Mg(^{3}Pu) + S(^{3}Pg)	2.71	—

configurations, $7\sigma^2\,8\sigma^2\,3\pi^4$ and $7\sigma^2\,8\sigma^1\,9\sigma^1\,3\pi^4$ [9–14]. The energies of the vibrational and rotational levels are obtained when solving the radial equation; the Numerov and Cooley methods [5] were used for that (see Table 3.2). Table 3.1 gives the spectroscopic constants of the bound electronic states of MgS, and we provide the equilibrium distances (bohr), the harmonic wave numbers, the anharmonic terms and the rotational constants and compare them to experimental values. Our results are very close to those of the literature and hence prove the reliability of our theoretical calculations. The theoretical vibrational spectrum could be mapped using the MRCI PECs. Figure 3.3 presents the spectrum obtained by using the transition moments between the ground state $X^1\Sigma^+$ and $1^1\Pi$, $2^1\Pi$ states (the transition moments between sigma states could not be calculated). According to this spectrum, the most probable transition is for $1^1\Pi$ state, the energy related for this transition is between $3.89 \times 10^3\,cm^{-1}$ and $15.79 \times 10^3\,cm^{-1}$. The FC factors for the different transitions between the vibrational levels of ground state and those of the $^1\Sigma^+$ and $^1\Pi$ bound states have been also used to simulate the vibrational spectrum (Figure 3.4). The transitions were fitted by Gaussian functions, and the line intensities were estimated from the FC factors by adopting a Boltzmann distribution to simulate the population of each rotational level at 300K. The spectrum depicted is dominated by transitions involving the ground state $X^1\Sigma^+$ to $1^1\Pi$, $2^1\Sigma^-$, $2\,^1\Pi$, but the most significant transition is for $2^1\Sigma^+$ vibrational levels (the most intense peak), and the energy related for this is between $2.10^3\,cm^{-1}$ and $10^4\,cm^{-1}$ .The equilibrium distance Re ($X^1\Sigma^+$) = 4.103 bohr is relatively close to Re ($2^1\Sigma^+$) = 4.20 bohr; hence the most probable transition is (v', v'') = (0,0).

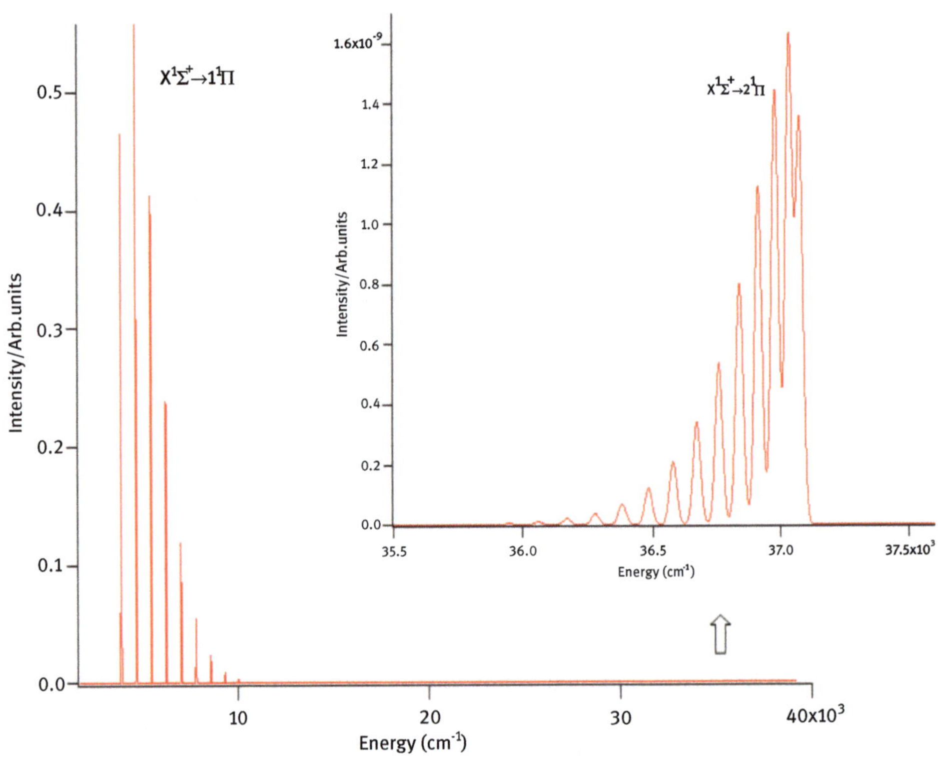

Figure 3.3: The transition momenta based simulated vibrational spectrum of MgS.

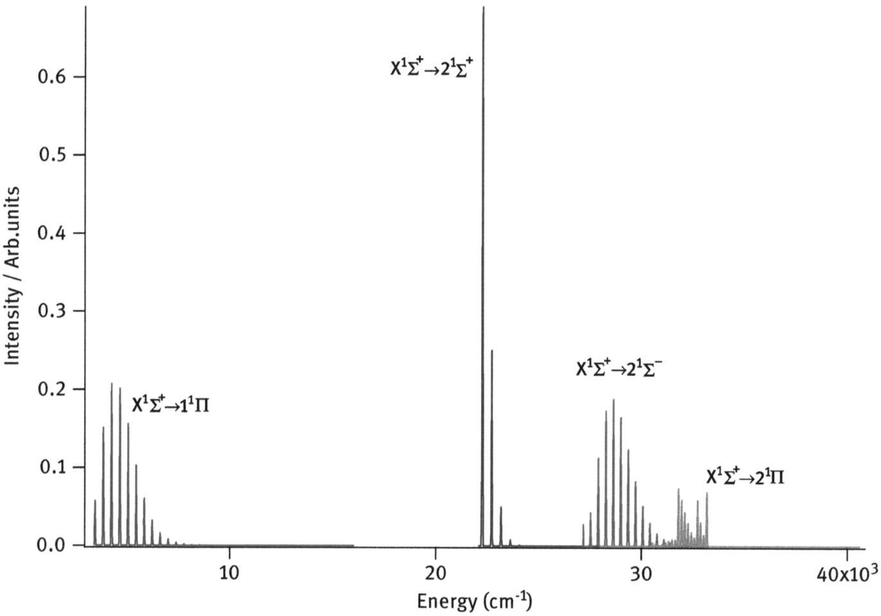

Figure 3.4: F.C factors based spectrum. FC factors versus ΔE of MgS.

3.3.2 MgS^{2+} dication

According to our theoretical calculations, the ground state $\mathbf{X^3\Sigma^-}$ of the MgS^{2+} dication has the following electronic configuration $(7\sigma)^2(8\sigma)^2(3\pi)^2$ obtained after the removal of the two electrons from the outermost 3π-orbital of the neutral molecule $(7\sigma)^2(8\sigma)^2(3\pi)^4$ which is of $\mathbf{X^1\Sigma^+}$ symmetry [9]. The equilibrium distance of the ground state of the dication is Re = 4.8 bohr, and this state presents a relatively deep potential well to the dissociation channel. The dissociation energy found is 1.20 eV (see Table 3.5). Figure 3.5 depicts the MRCI potential energy curves of singlet, triplet and quintet lowest states correlated to the six lowest dissociation asymptotes: **Mg^+ (^2S_g) + S^+ (^4S_u), Mg^+ (^2S_g) + S^+ (^2D_u), Mg^+(^2S_g) + S^+ (^2P_u), Mg^+ (^2P_u) + S^+ (^4S_u), Mg^+ (^2P_u) + S^+ (^2D_u)** and **Mg^+ (^2P_u) + S^+ (^2P_u)**. The energy ordering of the asymptotes is deduced from the literature [8] and compared to the values from our calculations; they are found in good agreement for the first four asymptotes and with a little difference for the Mg^+ (^2P_u) + S^+ (^2D_u) and Mg^+ (^2P_u) + S^+ (^2P_u) asymptotes (see Table 3.4). Table 3.5 lists the dominant configurations in the wave functions of the electronic states quoted at the equilibrium distance of the ground state $\boldsymbol{X^3\Sigma^-}$ (4.8 bohr). The five electronic bound states are $\boldsymbol{X^3\Sigma^-}$, $\mathbf{1^1\Delta}$, $\mathbf{1^3\Pi}$, $\mathbf{1^1\Sigma^+}$, $\mathbf{1^1\Pi}$ three states with very shallow potential well $\mathbf{2^3\Sigma^-}$, $\mathbf{2^1\Delta}$ **and** $\mathbf{2^1\Sigma^+}$, 11 electronic dissociative states $\mathbf{1^5\Sigma^-}$, $\mathbf{3^3\Sigma^-}$, $\mathbf{1^3\Delta}$, $\mathbf{2^3\Delta}$, $\mathbf{3^3\Delta}$, $\mathbf{4^1\Delta}$, $\mathbf{1^3\Sigma^+}$, $\mathbf{2^3\Sigma^+}$, $\mathbf{3^1\Pi}$, $\mathbf{2^3\Pi}$, $\mathbf{3^3\Pi}$ are also shown in Figure 3.5. On the other hand, some crossings of the PECs could be observed: **$1^1\Delta$ with $1^3\Pi$, $1^1\Sigma^+$ with $1^1\Pi$, $2^3\Sigma^-$with $1^3\Pi$** but the more

Table 3.4: Energies of the four lowest dissociation limits of MgS^{2+}.

Asymptotes	Values from Nist (eV)	MRCI/AVZ (eV)
$Mg(^1Sg)+S(^3Pg)$	0.00	0.00
$Mg(^1Sg)+S(^1Dg)$	1.15	1.36
$Mg(^3Pu)+S(^3Pg)$	2.71	–

Table 3.5: Equilibrium distances (R_e, in bohr) and dominant electronic configurations of the MgS^{2+} electronic states quoted for $R = 4.8$ bohr, correlating to the five lowest dissociation limits. T_e (in eV) corresponds to the MRCI adiabatic excitation energy. We give, between brackets, the leading CI coefficient in the CASSCF wavefunction. D_e (in eV) is calculated as the difference between the energy of the electronic state at equilibrium and the dissociation energy.

State	Electron configurations	R_e	T_e	D_e
$X^3\Sigma^-$	$(7\sigma)^2(8\sigma)^2(3\pi)^2$ (0.924) and $(7\sigma)^2(8\sigma)^1(9\sigma)^1(3\pi)^2$ (0.192)	4.76	00.00	1.20
$1^1\Delta$	$(7\sigma)^{2(}8\sigma)^2(3\pi)^2$ (0.681) and $(7\sigma)^{2(}8\sigma)^1(9\sigma)^1(3\pi)^2$ (0.099)	4.74	01.11	1.69
$1^3\Pi$	$(7\sigma)^2(8\sigma)^1(3\pi)^3$ (0.962) and $(7\sigma)^1(8\sigma)^2(3\pi)^3$ (0.069)	4.80	01.48	1.09
$1^1\Sigma^+$	$(7\sigma)^2(8\sigma)^2(3\pi)^2$ (0.651) and $(7\sigma)^2(8\sigma)^2(3\pi^*)^2$(0.651)	4.71	01.99	1.62
$1^1\Pi$	$(7\sigma)^2(8\sigma)^1(3\pi)^3$ (0.660) and $(7\sigma)^2(8\sigma)^2(9\sigma)^1(3\pi)^1$ (0.065)	4.99	02.57	0.52

important crossings are found for the $\mathbf{1^5\Sigma^-}$, cutting all the other bound states; hence their ro-vibrational levels should be perturbed. The proximity of the electronic states induces interactions and numerous couplings between the angular momenta that produce avoided crossings between states of the same symmetry and spin multiplicity as **$1^3\Sigma^+$ and $2^3\Sigma^+$, $1^1\Delta$ and $2^1\Delta$, $2^3\Sigma^-$and $3^3\Sigma^-$, $2^3\Pi$ and $3^3\Pi$, $1^1\Sigma^+$ and $2^1\Sigma^+$, $2^3\Delta$ and $3^3\Delta$, $2^1\Pi$ and $3^1\Pi$;** in consequence, the analysis of the vibrational spectrum is complicated. The energies of the vibrational and rotational levels are obtained when solving the radial equation. The Numerov and Cooley methods [5] were used for that (Table 3.6). Table 3.7 gives the spectroscopic constants of the bound electronic states of MgS^{2+}. We provide the equilibrium distances (bohr), the harmonic wavenumbers (cm^{-1}), the anharmonic terms and the rotational constants (B_e and ω_{ex}). We have calculated the FC factors for the different transitions: from the vibrational levels of $X^1\Sigma^+$ of the neutral to other vibrational levels of the following states of the dication: $X^3\Sigma^-$, $1^1\Sigma^+$, $1^1\Pi$, $1^3\Pi$. FC factors have been used to simulate the vibrational double-ionization spectrum in order to obtain a fingerprint of this dication and to facilitate its detection and to help future experimental spectroscopic works on MgS^{2+} (Figure 3.6). Only transitions with DJ = 0 were considered. This approach, where the FC factors are calculated for the double-ionization spectrum simulation, was widely discussed and validated in similar works [15–17].

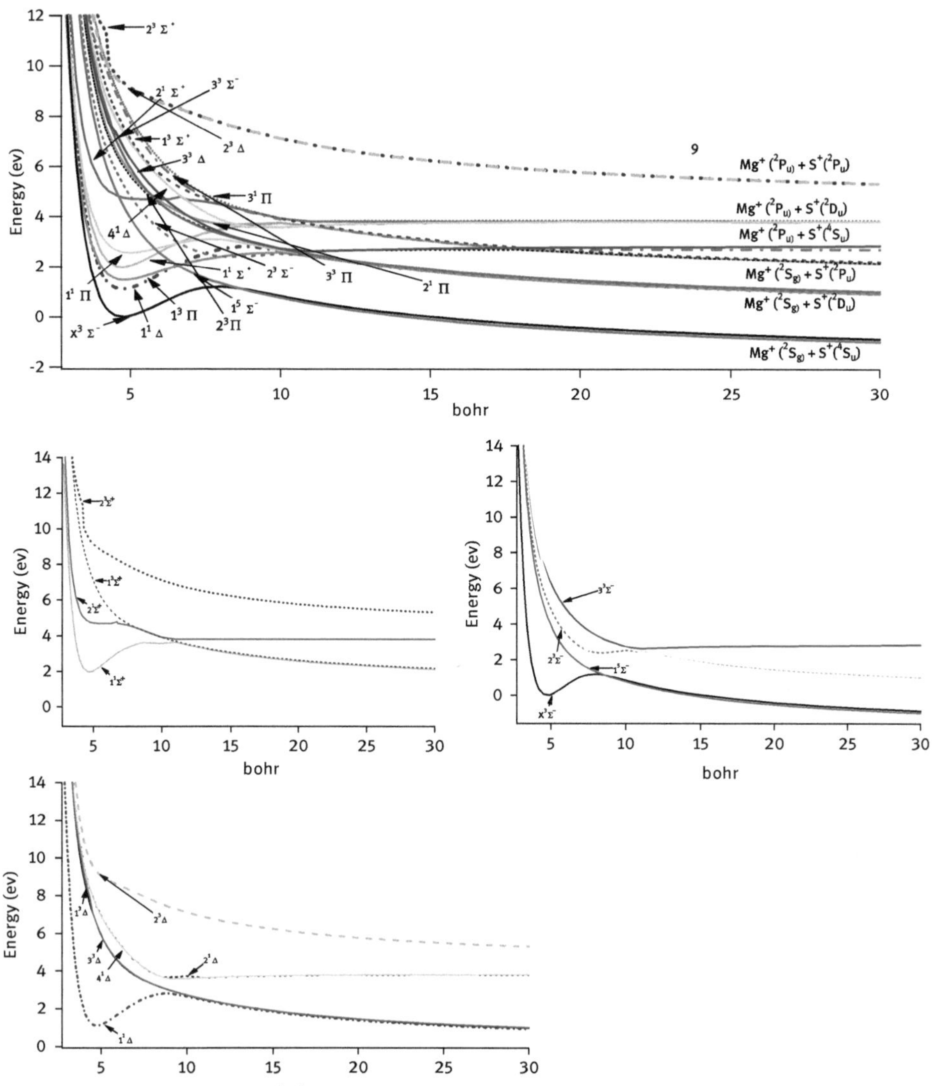

Figure 3.5: The potential energy curves of MgS^{2+} states (each symmetry is also represented separately).

The first "band" obtained corresponds to the transition:

$$MgS^{2+}\,(X^3\Sigma^-, \mathrm{v}=0) + 2e - - - - > MgS(X^1\Sigma^+, \mathrm{v}=0) + h\mathrm{v}.$$

This transition appears with a significant intensity due to favourable FC factors. The shapes of the potential energy wells of these electronic states are

Table 3.6: Vibrational levels of the different electronic states of MgS^{2+} and those of the ground state of MgS. All values are in cm^{-1}.

v	$^*X^1\Sigma^+$	$X^3\Sigma^-$	$1^3\Pi$	$1^1\Sigma^+$	$1^1\Pi$	$1^1\Delta$
0	255.620	160.485	132.860	167.278	113.190	166.473
1	763.507	481.207	395.636	499.072	336.993	496.832
2	1,266.359	801.306	654.676	827.309	557.556	823.914
3	1,763.850	1,117.589	910.015	1,152.005	774.883	1,147.743
4	2,255.954	1,429.545	1,161.685	1,473.159	988.976	1,468.309
5	2,742.721	1,736.411	1,409.708	1,790.791	1,199.829	1,785.616
6	3,224.174	2,038.332	1,654.114	2,104.917	1,407.396	2,099.696
7	3,700.270	2,335.739	1,894.919	2,415.539	1,611.643	2,410.585
8	4,171.057	2,629.233	2,132.142	2,722.666	1,812.559	2,718.305
9	4,636.688	2,919.336	2,365.798	3,026.305	2,010.092	3,022.862
10	5,097.430	3,206.407	2,595.904	3,326.464	2,204.188	3,324.262

* Ground state of MgS.

Table 3.7: MRCI spectroscopic constants of the ground electronic state of MgS and those of the electronic states of MgS^{2+} including the harmonic wavenumber (ω_e), the anharmonic terms ($\omega_e x_e$, $\omega_e y_e$), the rotational constants (β_e, α_e). All values are in cm^{-1}.

State	ω_e	ω_{exe}	ω_{eye}	β_e	α_e
MgS					
$X^1\Sigma^+$	512.608	-2.2747	−0.05438	0.25868	0.00142
MgS^{2+}					
$X^3\Sigma^-$	318.282	2.08491	-0.05324	0.19177	0.00179
$1^3\Pi$	266.544	−1.89341	0.00570	0.18866	0.00190
$1^1\Sigma^+$	335.367	−1.79095	0.00273	0.19606	0.00147
$1^1\Pi$	227.044	−1.65612	0.00396	0.17481	0.00186
$1^1\Delta$	333.658	-1.65612	0.00396	0.19342	0.00141

in favour of a good overlap of the vibrational levels. The spectrum is represented in Figure 3.6 and an assignment of the transitions is given in agreement with the experimental mass spectrum of Figure 3.1. All the transitions have a significant intensity except for $1^1\Pi$. An explanation is that the Re ($1^1\Pi$) = 4.9 bohr, very different from Re ($X^1\Sigma^+$) = 4.10 bohr of the neutral ground state; consequently, the transitions are between vibrational states of $X^1\Sigma^+$ and those of the continuum of $1^1\Pi$ with unfavourable overlap (see Figure 3.6), leading to weak intensities. The four peaks of the experimental spectrum are so assigned; the four electronic states of MgS^{2+} involved are hence $X^3\Sigma^-$, $1^1\Sigma^+$, $1^1\Pi$ and $1^3\Pi$.

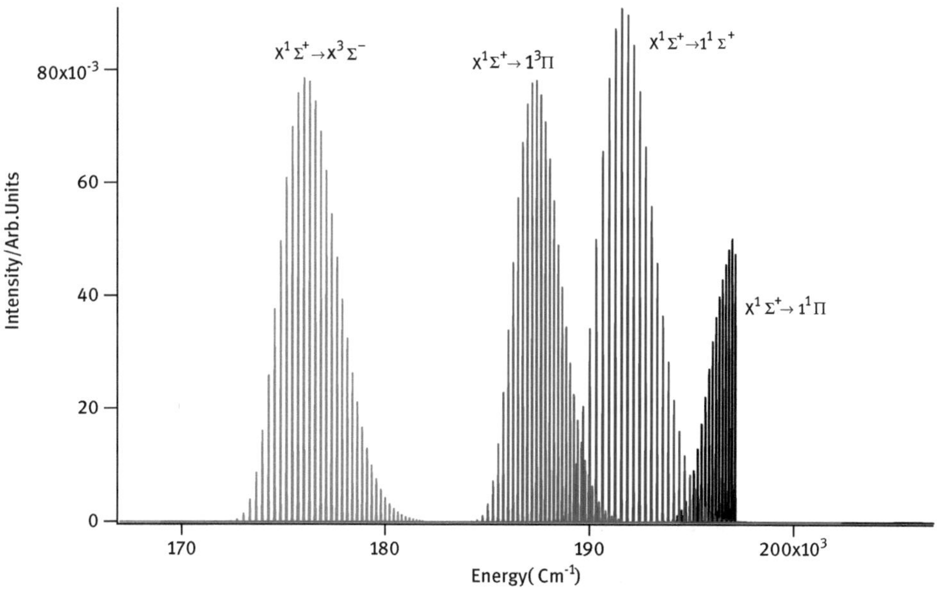

Figure 3.6: FC factors based on double-ionization simulated vibrational spectrum. FC factors versus ΔE of MgS^{2+}.

3.3.3 SiN^{2+} dication study

3.3.3.1 Determination of the low-lying electronic states of SiN^{2+}

The low-lying doublet and quartet electronic states of SiN^{2+} and its neutral parent molecule computed at the MRCI/aug-ccpV5Z level are mapped Figures 3.7 and 3.8, respectively. The minimum of the electronic ground state $X^2\Sigma^+$ of the neutral SiN is taken as the reference energy. All the electronic states correlating to the four lowest dissociation limits (i. e. $Si^+(^2P_u) + N^+ (^3P_g)$, $Si^{2+} (^1S_g) + N (^4S_u)$, $Si^+ (^2P_u) + N^+ (^1D_g)$ and $Si^+ (^2P_u) + N^+ (^1S_g)$) are calculated. The energy of these asymptotes and the correlated electronic states are given in Table 3.8. These asymptotes are not represented in the figures for more clarity. The $X^4\Sigma^-$ electronic ground state of SiN^{2+} corresponds to the removal of two electrons from the 2π MOs of the electronic ground state of the parent neutral molecule. Its potential energy curve shows a well of about 1.40 eV. The adiabatic double-ionization energy of SiN is found at 18.45eV, calculated as the energy difference between the minima of the ground states of the neutral and that of the dication. At least, six electronic excited states of SiN^{2+} are bound and show potential wells of several meV. In Table 3.9, the dominant electronic configurations of these states, together with the equilibrium distances, are given. The respective dissociation limits of the excited states are found lower in energy than their potential wells, proving the metastable character of these states. At 35 eV above the neutral ground state, the calculated electronic states are close in energy and can lead to

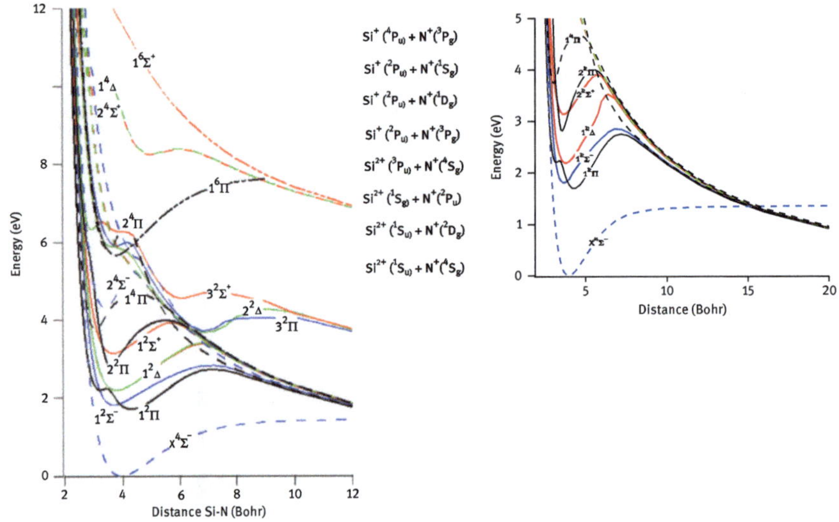

Figure 3.7: MRCI/aug-cc-pV5Z potential energy curves of the lowest states of SiN^{2+}. (short (left panel) and long range (right panel).

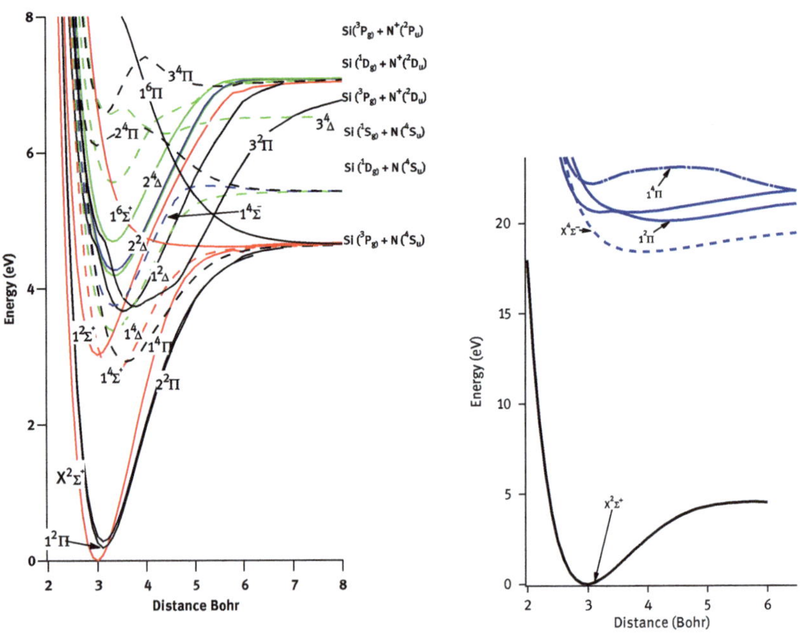

Figure 3.8: Potential energy curves of the electronic states of SiN (left panel) and ground states of the SiN and SiN^{2+} species (right panel) with the close lying doublet and quartet states of the dication.

Table 3.8: Energies of the lowest dissociation limits of SiN^{2+}, evaluated using the dissociation energy of the electronic ground state of the neutral SiN and an adequate thermo-chemistry cycle. The nature of the resulting electronic states is also reported.

Asymptote	Electronic states	Energy (eV)
$Si(^3P_g) + N\ (^4S_u)$	$^2\Pi$, $^2\Sigma^+$, $^4\Pi$, $^4\Sigma^+$, $^6\Sigma^+$, $^6\Pi$	0
$Si^{2+}(^1S_g) + N\ (^4S_u)$	$^4\Sigma^-(2)$, $^2\Pi(2)$, $^2\Sigma^-$, $^2\Delta$, $^4\Pi(2)$, $^2\Sigma^+$, $^4\Delta(3)$	16.346
$Si^{2+}\ (^1S_g) + N\ (^2D_u)$	$^2\Sigma^-$, $^2\Delta$, $^2\Pi$, $^2\Sigma^+$	18.729
$Si^{2+}\ (^3P_u) + N\ (^4S_g)$	$^6\Sigma^+$, $^4\Delta$, $^4\Sigma^+$, $^6\Pi$	22.883

Table 3.9: Dominant configurations and equilibrium distances (R_e *in* bohr) of the bound electronic states of SiN^{2+}.

State	Electron configuration	Coefficient	R_e(bohr)
SiN^{2+}			
$X^4\ \Sigma^-$	$5\sigma^2\ 6\sigma^2\ 2\pi^1_x\ 2\pi^2_y$	(0.7773)	4.0
$1^{\pi}\pi$	$5\sigma^2\ 6\sigma^2\ 7\sigma^1\ 2\pi^1_x\ 2\pi^1_y$	(0.7836)	4.4
$1^2\Sigma^-$	$5\sigma^2\ 6\sigma^2\ 7\sigma^1\ 2\pi^2_y$	(0.6918)	3.7
$1^2\Sigma^+$	$5\sigma^2\ 6\sigma^2\ 7\sigma^1\ 2\pi^2_x$	(0.6749)	3.7
$1^4\Pi$	$5\sigma^2\ 6\sigma^1\ 7\sigma^1\ 2\pi^2_x\ 2\pi^1_y$	(0.8263)	3.1
	$5\sigma^2\ 6\sigma^2\ 2\pi^2_x\ 2\pi^1_y$	(0.3199)	
$1^6\Pi$	$5\sigma^2\ 6\sigma^1\ 7\sigma^1\ 2\pi^2_x\ 2\pi^1_y$	(0.9596)	3.8
SiN			
$X^2\ \Sigma^+$	$5\sigma^2\ 6\sigma^2\ 7\sigma^1\ 2\pi^2_x\ 2\pi^2_y$	(0.9126)	3.0
$1^2\Pi$	$5\sigma^2\ 6\sigma^2\ 7\sigma^2\ 2\pi^1_x\ 2\pi^2_y$	(0.9263)	3.2
$1^2\ \Sigma^-$	$5\sigma^2\ 6\sigma^2\ 7\sigma^2\ 2\pi^2_x\ 2\pi^1_y$	(0.9079)	3.4
$1^4\Pi$	$5\sigma^2\ 6\sigma^2\ 7\sigma^1\ 2\pi^2_x\ 2\pi^2_y$	(0.6463)	4.0
$1^4\Sigma^+$	$5\sigma^2\ 6\sigma^2\ 7\sigma^1\ 2\pi^2_x\ 2\pi^2_y$	(0.6594)	3.2
$1^6\Pi$	$5\sigma^2\ 6\sigma^2\ 7\sigma^1\ 2\pi^2_x\ 2\pi^2_y$	(0.9719)	3.6

interactions through vibronic or spin–orbit couplings and to the mixing of their electronic wave functions. Figure 3.7 shows the avoided crossings between the $1^2\Pi$ and the $2^2\Pi$ for R_{SiN} at about 3.6 bohr, between $1^2\Sigma^+$ and $2^2\Sigma^+$ at 5.8 bohr and also between the $1^2\Delta$ and the $2^2\Delta$ for R_{SiN} at more than 6 bohr. These avoided crossings lead to the formation of local minima in the upper states.

Table 3.10 gives the equilibrium distances (R_e in bohr), the rotational constants (B_e in cm^{-1}), the harmonic wavenumbers (ω_e in cm^{-1}) and the anharmonic term ($\omega_e x_e$ in cm^{-1}) deduced from the MRCI potential energy curves and the resolution of the nuclear motion calculations.

As no experimental data are known for this species, all these values are predictive. For the $X^4\Sigma^-$ electronic ground state of this dication, the equilibrium distance is found at 3.936 bohr. This value is clearly larger than the one

Table 3.10: Main spectroscopic constants including the equilibrium distances (R_e in bohr), the equilibrium rotational constant (B_e in cm^{-1}), αe, γ; the harmonic and anharmonic wave numbers (ω_e and $\omega_e x_e$ in cm^{-1}) of the bound electronic states of SiN^{2+} calculated at the MRCI/AV5Z level.

State	ω_e	$\omega_e xe$	ω_{eye}	B_e	α_e	γ	R_e	D_e
$X^4\Sigma^-$	390.3	−1.510	−0.13217	0.41605	0.00645	0.00024	3.936	1.432
$1^4\Pi$	997.2	−14.378	−1.39137	0.68394	0.00812	−0.00077	3.070	1.855
$2^2\Sigma^+$	456.6	−6.879	0.13031	0.47889	0.00893	0.00012	3.669	
$1^2\Delta$	415.9	−4.263	0.09762	0.45112	0.00753	0.00014	3.779	
$1^2\Pi$	367.1	−2.658	−0.00345	0.34826	0.00309	0.00002	4.302	
$1^2\Sigma^-$	434.6	−8.473	0.16429	0.47687	0.01047	0.00007	3.677	
$1^6\Pi$	457.7	44.967	−3.71664	0.59855	0.00397	0.00299	3.282	7.528

computed at 2.985 bohr for the electronic ground state $X^2\Sigma^+$ of the neutral SiN, resulting in the depletion of the 2π bonding MOs. As the bond length of the dication is longer than that of the neutral, the equilibrium rotational constant of the electronic ground state of the dication is computed at 0.41605 cm^{-1}, smaller than the one of 0.72305 cm^{-1} computed for the ground state of the neutral molecule. Concerning the vibration constants of the dication, the ground state values of ω_e and $\omega_e x_e$ are computed as 390.305 cm^{-1} and 1.51086 cm^{-1}, respectively. For the neutral molecule, these values are computed at 1142.82 cm^{-1} and 6.37467 cm^{-1}, respectively. All the values computed for the $X^2\Sigma^+$ state of the neutral molecule are in close agreement with the experimental determination (R_e = 2.985 bohr, B_e = 0.723 cm^{-1}, ω_e= 1142.82 cm^{-1} and $w_e x_e$= 6.37 cm^{-1} Table 3.11). A similar accuracy is expected for the values predicted for the electronic states of the dication.

3.3.3.2 Simulation of the double-ionization spectrum of SiN

To obtain a fingerprint of this dication in order to facilitate its detection and to help future experimental spectroscopic works on SiN^{2+}, the double-ionization spectrum of SiN (Figure 3.9) has been simulated. The FC factors between the electronic ground state $X^2\Sigma^+$ of the neutral SiN molecule and the electronic states of SiN^{2+} have been computed using the LEVEL program [7]. Only transitions with DJ = 0 were considered. The first “band” obtained corresponds to the transition:

$$SiN^{2+}(X^4\Sigma^+, v=0) + 2e^- - - - - - - - - >, SiN(X^2\Sigma^+, v=0) + hv.$$

This transition appears with a low intensity due to unfavourable FC factors. It could not be represented in our theoretical spectrum. The deepness and

Table 3.11: Main spectroscopic constants including the equilibrium distances, the equilibrium rotational constants (B_e in cm^{-1}), the rotational constants (β_e, α_e, γ). All values are in cm^{-1}, equilibrium distances (R_e, in bohr), T_e (in eV) corresponds to the MRCI adiabatic excitation energy, D_e (in eV) is calculated as the difference between the energy of the electronic state at equilibrium and the energy at the maximum of the potential barrier, the harmonic and anharmonic wave numbers (ω_e and ω_{exe}, ω_{eye} in cm^{-1}) of the bound electronic states of SiN.

	ω_e	ω_{exe}	ω_{eye}	B_e	α_e	R_e	T_e	D_e
$X^2\Sigma^+$	1142.8	6.37	-0.001	0.723	0.005	2.985	0	5.10
MRCI[18]	1124	7.0				3.011		4.26
SDCI[19]	1338					2.881		
SCF[20]	1138					2.964		
LDA, large ANO basis set[21]	1189	6.47				2.964		
MRCI/cc-pVTZ[22]	1129	6.37				3.007		
Ref[23][d]	1155					2.972		
Exp[24]	1151							4.58
Exp[25]	1151	6.46						
Exp[26]								4.68
Exp[27]	1151	6.47				2.972		
MRCI[18]	1151.4	4.47		0.731	0.006	2.968		
$1^2\pi$	1027.8	5.92	0.012	0.664	0.005	3.115	0.22	4.50
$2^2\Sigma^+$	1042.6	15.34	0.005	0.7π3	0.005	2.985		4.25
MRCI[18]	958	14.6				3.047		3.81
Exp[24]	1031	16.9						
Exp[26]	1031	16.9						
Exp[27]	1031	16.9				2.987		
$1^2\Sigma^-$	742.6	4.64	0.012	0.579	0.006	3.335	4.10	3.16
$1^2\pi$	775.1	6.07	0.069	0.586	0.006	3.316	4.32	2.78
$1^4\Sigma^+$	775.9	6.47	0.013	0.586	0.006	3.316	2.69	2.71
$1^4\Delta$	777.2	5.71	0.020	0.587	0.006	3.316	3.27	2.14
$1^4\pi$	639.4	π.79	0.208	0.505	0.005	3.571	2.85	1.83
$1^4\Sigma^-$	763.3	5.43	0.034	0.583	0.006	3.323	3.30	2.10
$2^4\Delta$	777.1	5.69	0.019	0.586	0.006	3.314	3.26	0.96
$1^6\pi$	988.5	-10.77	-0.516	0.506	0.003	3.567	8.42	2.70

[a] "best estimate" see original paper for details.

width of the potential well of this electronic state of the dication and the difference of equilibrium distance between the ground states of the neutral and the dication are not in favour of this transition. Similar situation is obtained for the excited states of the dication possessing low barrier heights and larger equilibrium bond distances compared to the electronic ground state of the neutral SiN (i. e. the $1^2\Pi$, $2^2\Sigma^+$, $1^2\Sigma^-$, $4^2\Pi$, $1^4\Sigma^+$, $1^4\Delta$, $1^4\Pi$, $1^4\Sigma^-$ and $2^4\Delta$). The only two allowed transitions, with favoured FC factors, spin and symmetry, are between doublets. The transition $SiN^{2+}(2^2\Sigma^+)$ $SiN(X^2\Sigma^+)$ is the most intense. The spectrum is represented in Figure 3.9 and an assignment of the

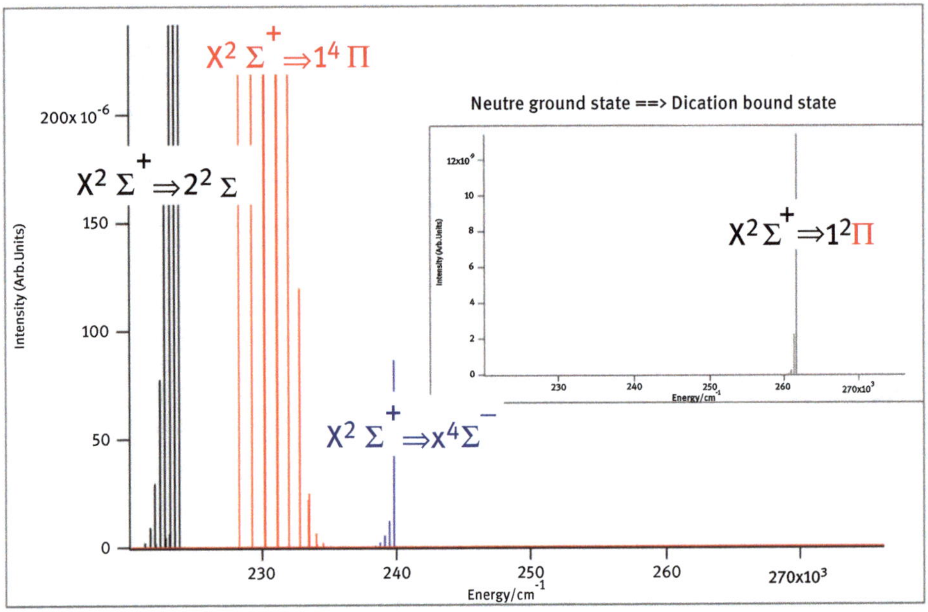

Figure 3.9: Franck–Condon factor-based simulation of the double-ionization spectrum of SiN.

transitions is given in agreement with the experimental mass spectrum of Figure 3.10. We should say that we have computed two transitions with favoured FC factors that are:

$$SiN^{2+}(1^4\Sigma^-, v=0) + 2e - - - - - > SiN(X^2\Sigma^+, v=0) \text{ and}$$

$$SiN^{2+}(1^4\Pi, v=0) + 2e - - - - - > SiN(X^2\Sigma^+, v=0).$$

As these transitions are not spin or symmetry allowed, they are not observed in the experimental mass spectrum.

3.4 Conclusion

In the present theoretical work, a large study and a global view of the potential energy curves, transition moment functions, dissociation energies and the spectroscopic constants of the bound electronic states of MgS and SiN and their respective dications MgS^{2+} and SiN^{2+} have been performed. Electronic ground states and excited electronic states have been accurately described here. Our calculations predict the existence of metastable electronic states of MgS^{2+} and SiN^{2+}. Most lines of the

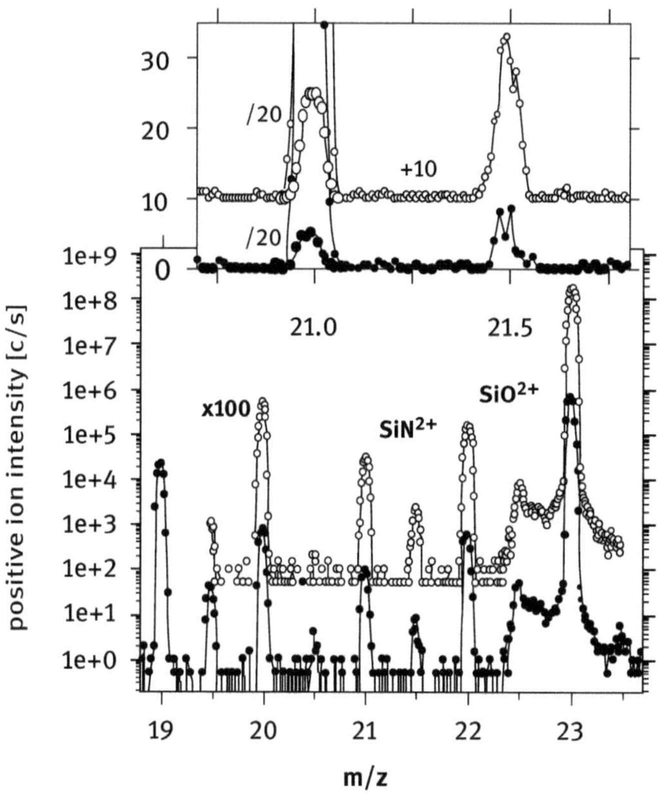

Figure 3.10: Experimental mass spectrum from reference [1] with the permission of the authors.

experimental mass spectra of these systems could be so assigned. The electronic states of the MgS^{2+} and SiN^{2+} dications have been determined using highly accurate ab initio computations. The Potential Energy Curves (PECs) were calculated with significant potential wells proving the possible existence of such species. The electronic ground state of SiN^{2+} is found of $X^4\Sigma^-$ nature ($X^3\Sigma^-$ for MgS^{2+}) and presents a potential well of about 1.40eV (1.20 eV for MgS^{2+}). Other excited electronic states are also predicted to be bound, and several couplings between these states are supposed to occur. The rotational and the vibrational spectroscopic constants of the bound states are computed. The FC double-ionization spectra of SiN and MgS are simulated using the potential energy curves of the bound states of the dications and those of the electronic ground states of the neutral molecules. The SiN spectrum is mainly composed by the contribution of two excited states of the dication (the $4^2\Pi$ and $2^2\Sigma^+$), whereas the peaks due to the ground state and the other bound excited states appear with relatively low intensities, because of the large difference of the equilibrium bond length of these states and that of the ground state of the neutral molecule. The only allowed transitions are first between the $X^2\Sigma^+$ state of the neutral and the $1^2\Sigma^+$ state of the dication, and second, much less intense, between the $X^2\Sigma^+$ state of the neutral and the $1^2\Pi$ state of the dication. The other double-ionization transitions are spin or symmetry forbidden. This simulated spectrum (Figure 3.9) is in agreement with the experimental mass spectrum produced at Arizona State University by Franzreb and Williams [1] (Figure 3.10). On the other hand, similarly, we could assign the four peaks of the experimental double-ionization spectrum of MgS. They are due to the transitions between the ground state of the neutral $X^1\Sigma^+$ and the $X^3\Sigma^-$, $1^1\Sigma^+$, $1^1\Pi$, $1^3\Pi$ states of the dication. In this work, we could assign the observed bands in the experimental mass spectra of SiN and MgS. All the data given here could also be helpful to future experimental works dealing with the spectroscopy of these dications.

Funding: This research was supported by the Moroccan Research Program under the reference grant n° **PU-SCH09/09** of the Ministry of Higher Education and a Marie Curie International Research Staff Exchange Scheme Fellowship within the 7th European Community Framework Program under grant n° **PIRSES-GA-2012-31754**.

We thank Klaus Franzreb and Peter Williams for their permission to include their unpublished experimental data in our theoretical work.

References

[1] Franzreb K, Williams P.. Experimental results, time period of 2004–2009 (unpublished). XXXX and http://www.rsc.org/suppdata/cp/c1/c1cp21566c/c1cp21566c.pdf.

[2] Ben Yaghlane S, Jaidane NE, Franzreb K, Hochlaf MA. Theoretical and experimental investigation of the diatomic dication SiO^{2+}. Chem Phys Lett. 2010;486:16 Experimental spectrum therein.

[3] Werner H.-J, PJ Knowles. An efficient internally contracted multiconfiguration–reference configuration interaction method. J Chem Phys. 1988;89:5803–5814 And references therein.

[4] Werner H-J, Knowles PJ, Knizia G, Manby FR, Schütz M, Celani P, Korona T, Lindh R, Mitrushenkov A, Rauhut G, Shamasundar KR, Adler TB, Amos RD, Bernhardsson A, Berning A, Cooper DL, Deegan MJO, Dobbyn AJ, Eckert F, Goll E, Hampel C, Hesselmann A, Hetzer G, Hrenar T, Jansen G, Köppl C, Liu Y, Lloyd AW, Mata RA, May AJ, McNicholas SJ, Meyer W, Mura ME, Nicklaß A, O'Neill DP, Palmieri P, Pflüger K, Pitzer R, Reiher M, Shiozaki T, Stoll H, Stone AJ, Tarroni R, Thorsteinsson T, Wang M, Wolf A. MOLPRO is a package of ab initio programs written by. Further details at http://www.molpro.net, XXXX.

[5] Numerov B. Publs. Observatoire Central Astrophys Russ. 1933;2:188.

[6] Cooley JW. An improved eigenvalue corrector formula for solving the Schrödinger equation for central fields. Math Computation. 1961;15:363–374.

[7] Le Roy RJ. Level 7.2. University of Waterloo Chemical Physics Research Report 2002. 642.

[8] Linstrom PJ, Mallard WG. Standard Reference Database Number 69. http://webbook.nist.gov.

[9] Ribas VW, Ferrao LFA, Neto OR, FBC Machado. Transition probabilities and molecular constants of the low-lying electronic states of the MgS molecule. Chem Phys Lett. 2010;492:19–24.

[10] Maracano M, Barrow RF. Rotational analysis of a $^1\Sigma^{+-}\,^1\sigma^+$ system of gaseous SrS. Trans Faraday Soc. 1970;66:1917–1919.

[11] Walker KA, Gerry MCL. Microwave Fourier transform spectroscopy of magnesium sulfide produced by Laser Ablation. J Mol Spectr. 1997;182:178–183.
[12] Partridge H, Langhoff SR, CW Bauschlicher. Theoretical study of the alkali and alkaline-earth monosulfides. J Chem Phys. 1988;88:6431.
[13] Chase W, Curnutt JL, Downey JR, McDonald RA, Syverud AN. JANAF thermochemical tables. 1982 Supplement. J Phys Chem Reference Data. 1982;11:695–940.
[14] Chambaud G, Guitou M, Hayashi S. Specific electronic properties of metallic molecules MX correlated to piezoelectric properties of solids MX. Chem Phys. 2008;352:147–156.
[15] Bennett R, ADJ Critchley, GC King, RJ LeRoy, IR McNab. Interpreting vibrationally resolved spectra of molecular dications. Mol Phys. 1999;97:35–42.
[16] Yencha J, AM Juarez, SP Lee, GC King, FR Bennett, Kemp F, IR McNab. Photo-double ionization of hydrogen iodide: experiment and theory. Chem Phys. 2004;303:179–187.
[17] Brites V, Hammoutène D, Hochlaf M. Spectroscopy, metastability, and single and double ionization of AlCl. J Phys Chem A. 2008;112:13419–13426.
[18] Borin AC. A complete active space self-consistent field and multireference configuration interaction analysis of the SiN B $^2\Sigma^+$-X $^2\Sigma^+$ transition moment. Chem Phys Lett. 1996;262: 80–86.
[19] Muller-Plathe F, Laaksonen L. Hartree–Fock-limit properties for SiC, SiN, Si2, Si2* and SiS. Chem Physlett. 1989;160:175.
[20] McLean AD, Liu B, Chandler GS. Computed self-consistent field and singles and doubles configuration interaction spectroscopic data and dissociation energies for the diatomics B_2, C_2, N_2, O_2, F_2, CN, CP, CS, PN, SiC, SiN, SiO, SiP, and their ions. J Chem Phys. 1992;97:8459–8464.
[21] Chong DP. Local density studies of diatomic AB molecules, A, B, C, N, O, F, Si, P, S, and Cl. Chem Phys Lett. 1994;220:102.
[22] Cai ZL, Martin JML, JP Francois, Gijbels R. Ab initio study of the $X^2\Sigma^+$ and $A^2\Pi$ states of the SiN radical. Chem Phys Lett. 1996;252:398.
[23] Curtiss LA, Raghavachari K, Trucks GW, Pople JA. Gaussian2 theory for molecular energies of first and second row compounds. J. Chem Phys. 1992;94:931.
[24] Foster SC. The vibronic structure of the SiN radical. J Mol Spectrosc. 1989;137:430.
[25] Naulin C, Costes M, Moudden Z, Chanem N, Dorthe G. Measurements of the radiative lifetimes of MgO(B $^1\Sigma^+$, d $^3\Delta$, D $^1\Delta$) states. Chem Phys Lett. 1991;94:7221.
[26] Bredohl H, Dubois I, Houbrechts Y, Singh M. The emission spectrum of SiN. Can J Phys. 1976;54:680–688.
[27] Herzberg G. Molecular spectra and molecular structure. Vol. 1. Spectra of diatomic molecules. New York: Van Nostrand Reinhold, 1950.

Jayvant Patade and Sachin Bhalekar

4 Analytical Solution of Pantograph Equation with Incommensurate Delay

Abstract: Pantograph equation is a delay differential equation (DDE) arising in electrodynamics. This paper studies the pantograph equation with two delays. The existence, uniqueness, stability and convergence results for DDEs are presented. The series solution of the proposed equation is obtained by using Daftardar-Gejji and Jafari method and given in terms of a special function. This new special function has several properties and relations with other functions. Further, we generalize the proposed equation to fractional-order case and obtain its solution.

Keywords: pantograph equation, Daftardar-Gejji and Jafari method, proportional delay

4.1 Introduction

Ordinary differential equations (ODE) are widely used by researchers to model various phenomena arising in Science and Technology. However, it is observed that such equations cannot model the actual behavior of the system. Since the ordinary derivative is a local operator, it cannot model the memory and hereditary properties in the real-life models. Such phenomena can be modeled in a more accurate way by introducing some nonlocal component, e.g. delay in it.

The delay differential equations (DDEs) are the equations where the rate of change of certain quantity depends on the value of that quantity at previous time [1]. The DDE models are proved very useful while modeling natural systems [2, 3]. The analysis of DDE is more difficult than that of ODE. The characteristic equation corresponding to DDE is a transcendental equation unlike a polynomial in case of ODE. Some first-order nonlinear DDEs may exhibit chaotic oscillations [4].

The DDE

$$y'(x) = ay(x) + by(qx), \tag{4.1}$$

is a famous equation called pantograph equation arising in electrodynamics. The pantograph [5, 6] is a device used in electric trains to collect electric current from the overload lines. The equation was modeled by Ockendon and Tayler [7] in 1971. The analytical solution of pantograph eq. (4.1) and its asymptotic properties are discussed by Kato [8]. Liu [9] used trapezium rule to find numerical solution of eq. (4.1). The DJM is used by Bhalekar and Patade [10] to find analytical series solution of eq. (4.1). The

https://doi.org/10.1515/9783110467215-004

convergence of the series solution is discussed and the properties of the novel special function defined in terms of the series are also discussed by these authors. Iserles [11] presented generalization of pantograph equation namely

$$y'(t) = Ay(t) + By(qt) + Cy'(qt) \tag{4.2}$$

Bellen, Guglielmi and Torelli [12] studied the stability properties of θ-methods for eq. (4.2). The eq. (4.2) in complex plane is described in [13]. Koto [14] discussed the stability of Runge-Kutta methods for the eq. (4.2).

Yet another generalization namely multi-pantograph equation

$$y'(t) = \lambda y(t) + \sum_{i=1}^{l} \mu_i y(q_i t) \tag{4.3}$$

is proposed by Liu and Li [15]. Adomian decomposition method and variational iteration method are used to solve some particular cases of eq. (4.3) in [16] and [17] respectively. Runge-Kutta methods are used to solve this equation numerically by Li and Liu [18]. The approximate solution of eq. (4.3) with variable coefficients is presented in [19] in terms of Taylor polynomials. However, the solution of eq. (4.3) is not given in the literature in terms of a special function.

Our aim in this paper is to analyze the pantograph equation

$$y'(x) = ay(x) + by(px) + cy(qx) \tag{4.4}$$

with incommensurate delays. We use Daftardar-Gejji and Jafari method (DJM) to obtain series solution of eq. (4.4). We define the new special function using this series and analyze its properties. The paper is organized as below: Basic definition and results given in Section 4.2.1. The iterative scheme DJM is discussed in Section 4.2.2. Existence, uniqueness and convergence results are described in Section 4.2.3. Section 4.3 deals with stability analysis. The solution of pantograph equation is given in Section 4.4. Analysis of special function generated from this solution is given in Section 4.5. The equation is generalized to fractional-order case in Section 4.6. The conclusions are summarized in Section 4.7.

4.2 Preliminaries

4.2.1 Basic definitions and results

We recall some basic definitions and results from [20], [21] and [22] which will be used in this paper.

Definition 4.1.

The upper and lower incomplete gamma functions are defined as

$$\Gamma(n,x) = \int_x^{\infty} t^{n-1}e^{-t}dt \quad \text{and}$$

$$\gamma(n,x) = \int_0^{x} t^{n-1}e^{-t}dt \quad \text{respectively.}$$

Various properties of incomplete gamma functions are discussed in [20].

Definition 4.2.

Kummer's confluent hyper-geometric functions ${}_1F_1(a;c;x)$ *and* $U(a;c;x)$ *are defined as below*

$${}_1F_1(a;c;x) = \sum_{n=0}^{\infty} \frac{(a)_n}{(c)_n}\frac{x^n}{n!}, \quad c \neq 0,-1,-2,\cdots \text{and}$$

$$U(a;c;x) = \frac{\pi}{\sin(\pi c)}\left(\frac{{}_1F_1(a;c;x)}{\Gamma(c)\Gamma(1+a-c)} - x^{1-c}\frac{{}_1F_1(1+a-c;2-c;x)}{\Gamma(a)\Gamma(2-c)}\right),$$

$$-\pi < \arg x \leq \pi.$$

Definition 4.3.

The generalized Laguerre polynomials are defined as

$$\begin{aligned} L_n^{(\alpha)}(x) &= \sum_{m=0}^{n}(-1)^m \binom{n+\alpha}{n-m}\frac{x^m}{m!} \\ &= \binom{n+\alpha}{n}{}_1F_1(-n;\alpha+1;x). \end{aligned}$$

Definition 4.4.

A real function $f(x)$, $x>0$, is said to be in space C_α, $\alpha \in \mathbb{R}$, if there exists a real number $p(>\alpha)$, *such that* $f(x) = x^p f_1(x)$ where $f_1(x) \in C[0,\infty)$.

Definition 4.5.

A real function $f(x)$, $x>0$, *is said to be in space* C_α^m, $m \in \mathbb{N} \cup \{0\}$, if $f^{(m)} \in C_\alpha$.

Definition 4.6.

Let $f \in C_\alpha$ and $\alpha \geq -1$, *then the (left-sided) Riemann-Liouville integral of order* $\mu, \mu > 0$ *is given by*

$$I^\mu f(t) = \frac{1}{\Gamma(\mu)} \int_0^t (t-\tau)^{\mu-1} f(\tau) d\tau, \quad t > 0.$$

Definition 4.7.

The (left sided) Caputo fractional derivative of $f, f \in C_{-1}^m, m \in \mathbb{N} \cup \{0\}$, is defined as:

$$\begin{aligned} D^\mu f(t) &= \frac{d^m}{dt^m} f(t), \quad \mu = m \\ &= I^{m-\mu} \frac{d^m}{dt^m} f(t), \quad m-1 < \mu < m, \quad m \in \mathbb{N}. \end{aligned}$$

Note that for $0 \leq m-1 < \alpha \leq m$ and $\beta > -1$

$$\begin{aligned} I^\alpha (x-b)^\beta &= \frac{\Gamma(\beta+1)}{\Gamma(\beta+\alpha+1)} (x-b)^{\beta+\alpha}, \\ (I^\alpha D^\alpha f)(t) &= f(t) - \sum_{k=0}^{m-1} f^{(k)}(0) \frac{t^k}{k!}. \end{aligned}$$

Theorem 4.8.

Rudin [23] *Suppose* $\{f_n\}$ *is a sequence of functions defined on* E, and suppose that there exists $M_n \in \mathbb{R}$ *such that*

$$|f_n| \leq M_n, \quad \text{on } E, \quad n = 1, 2, 3 \cdots.$$

Then $\sum f_n$ *converges uniformly on* E *if* $\sum M_n$ *converges.*

4.2.2 Daftardar-Gejji and Jafari method

A new iterative method (DJM) was introduced by Daftardar-Gejji and Jafari [24] in 2006 for solving nonlinear functional equations. The DJM has been used to solve a variety of equations such as fractional differential equations [25], partial differential equations [26], boundary value problems [27, 28], evolution equations [29] and system of nonlinear functional equations [30]. The method is successfully employed to solve Newell-Whitehead-Segel equation, Fishers equation [31, 32], fractional-order logistic equation [33] and some nonlinear dynamical systems [34] also. Recently DJM has been used to generate new numerical methods [35–38] for solving differential equations. In this section we describe DJM which is very useful for solving the equations of the form

$$u = f + L(u) + N(u), \tag{4.5}$$

where f is a given function, L and N are linear and nonlinear operators respectively. DJM provides the solution to eq. (4.5) in the form of series

$$u = \sum_{i=0}^{\infty} u_i. \tag{4.6}$$

Since L is linear

$$L\left(\sum_{i=0}^{\infty} u_i\right) = \sum_{i=0}^{\infty} L(u_i). \tag{4.7}$$

The nonlinear operator N in eq. (4.5) is decomposed by Daftardar-Gejji and Jafari as below:

$$\begin{aligned} N\left(\sum_{i=0}^{\infty} u_i\right) &= N(u_0) + \sum_{i=1}^{\infty}\left\{N\left(\sum_{j=0}^{i} u_j\right) - N\left(\sum_{j=0}^{i-1} u_j\right)\right\} \\ &= \sum_{i=0}^{\infty} G_i, \end{aligned} \tag{4.8}$$

where

$$\begin{aligned} G_0 &= N(u_0) \quad \text{and} \\ G_i &= \left\{N\left(\sum_{j=0}^{i} u_j\right) - N\left(\sum_{j=0}^{i-1} u_j\right)\right\}, \quad i \geq 1. \end{aligned}$$

Using eqs (4.6), (4.7) and (4.8) in eq. (4.5), we get

$$\sum_{i=0}^{\infty} u_i = f + \sum_{i=0}^{\infty} L(u_i) + \sum_{i=0}^{\infty} G_i. \tag{4.9}$$

From eq. (4.9), the DJM series terms are generated as below:

$$\begin{aligned} u_0 &= f, \\ u_{m+1} &= L(u_m) + G_m, \quad m = 0, 1, 2, \cdots. \end{aligned}$$

The k-term approximate solution is given by

$$u = \sum_{i=0}^{k-1} u_i,$$

for suitable integer k.

Convergence of DJM is given in following results.

Theorem 4.9.

Bhalekar and Daftardar-Gejji [39] If N is $C^{(\infty)}$ in a neighborhood of y_0 and $\|N^{(n)}(y_0)\| \leq L$, for any n and for some real $L > 0$ and $\|y_i\| \leq M < \frac{1}{e}$, $i = 1, 2, \cdots$, then the series $\sum_{n=0}^{\infty} G_n$ is absolutely convergent to N and moreover,

$$\|G_n\| \leq LM^n e^{n-1}(e-1), \quad n = 1, 2, \cdots.$$

Theorem 4.10.

Bhalekar and Daftardar-Gejji [39] If N is $C^{(\infty)}$ and $\|N^{(n)}(y_0)\| \leq M \leq e^{-1}$, $\forall n$, then the series $\sum_{n=0}^{\infty} G_n$ is absolutely convergent to N.

4.2.3 Existence, uniqueness and convergence

In this section we generalize theorems described in [40]. The equation

$$y'(x) = f(x, y(x), y(px), y(qx)),$$

is a particular case of time-dependent DDE

$$y'(x) = f(x, y(x), y(x - \tau_1(x)), y(x - \tau_2(x))) \quad \text{with} \quad \tau_1(x) = (1-p)x, \quad \tau_2(x) = (1-q)x.$$

Theorem 4.11.

*(**Local existence**) Consider the equation*

$$\begin{aligned} y'(x) &= f(x, y(x), y(x - \tau_1(x)), y(x - \tau_2(x))), \quad x_0 \leq x < x_f, \\ y(x_0) &= y_0, \end{aligned} \tag{4.10}$$

and assume that the function $f(x, u, v, w)$ is continuous on $A \subseteq [x_0, x_f) \times \mathbb{R}^m \times \mathbb{R}^m \times \mathbb{R}^m$ and locally Lipschitz continuous with respect to u, v and w. Moreover, assume that the delay function $\tau_1(x) \geq 0$, $\tau_2(x) \geq 0$ is continuous in $[x_0, x_f)$, $\tau_1(x_0) = 0$, $\tau_2(x_0) = 0$ and, *for some* $\xi > 0, x - \tau_1(x) > x_0, x - \tau_2(x) > x_0$ *in the interval* $(x_0, x_0 + \xi]$. *Then the problem* (4.10) *has a unique solution in* $[x_0, x_0 + \delta)$ *for some* $\delta > 0$ *and this solution depends continuously on the initial data.*

Theorem 4.12.

(***Global existence***) *Under the hypothesis of Theorem* Theorem 11, *if the unique maximal solution of* (4.10) *is bounded, then it exist on the entire interval* $[x_0, x_f)$.

Now, we present convergence result motivated from Bhalekar and Patade [41] for DJM solution.

Theorem 4.13.

Let f *be a continuous function defined on a four-dimensional rectangle* $R = \{(x, y_1, y_2, y_3) | 0 \le x \le b, -\delta \le y_1 \le \delta, -\mu \le y_2 \le \mu, -\eta \le y_3 \le \eta\}$ *and* $| f(x, y_1, y_2, y_3) | \le M, \ \forall (x, y_1, y_2, y_3) \in R$. *Suppose that* f *satisfies Lipschitz type condition* $| f(x, y_1, y_2, y_3) - f(x, u_1, u_2, u_3) | \le L_1 | y_1 - u_1 | + L_2 | y_2 - u_2 | + L_3 | y_3 - u_3 |$. *Then the DJM series solution of the initial value problem (IVP),*

$$y'(x) = f(x, y(x), y(px), y(qx)), \ y(0) = 1, \ 0 < p < 1, 0 < q < 1, \tag{4.11}$$

converges uniformly in the interval [0,b].□

proof.

The equivalent integral equation of eq. (4.11) *is*

$$y(x) = 1 + \int_0^x f(t, y(t), y(pt), y(qt))dt.$$

Using DJM, we get

$$\begin{aligned} y_0(x) &= 1, \\ y_1(x) &= \int_0^x f(t, y_0(t), y_0(pt), y_0(qt))dt. \\ \Rightarrow | y_1(x) | &\le Mx. \end{aligned}$$

Since $p, q \in (0, 1)$, $\dfrac{b}{p} > b$ and $\dfrac{b}{q} > b$.

$$\Rightarrow | y_1(px) | \le Mpx, \quad \forall x \in [0, b].$$

Further,

$$\begin{aligned}
y_2(x) &= \int_0^x (f(t, y_1(t) + y_0(t), y_1(pt) + y_0(pt), y_1(qt) + y_0(qt)) \\
&\quad - f(t, y_0(t), y_0(pt), y_0(qt)))dt. \\
\Rightarrow |\, y_2(x) \,| &\leq \int_0^x (L_1 \,|\, y_1(x) \,| + L_2 \,|\, y_1(pt) \,| + L_3 \,|\, y_1(qt) \,|)dt \\
&\leq M(L_1 + L_2 p + L_3 q)\frac{x^2}{2!} \\
&\leq M(L_1 + L_2 + L_3)\frac{x^2}{2!}.
\end{aligned}$$

$$\begin{aligned}
\Rightarrow |\, y_2(px) \,| &\leq Mp^2(L_1 + L_2 p + L_3 q)\frac{x^2}{2!}, \quad x \in [0, b] \\
&\leq M(L_1 + L_2 + L_3)\frac{x^2}{2!} \\
\text{and} \quad |\, y_2(qx) \,| &\leq Mq^2(L_1 + L_2 p + L_3 q)\frac{x^2}{2!}, \quad x \in [0, b] \\
&\leq M(L_1 + L_2 + L_3)\frac{x^2}{2!}.
\end{aligned}$$

Similarly,

$$\begin{aligned}
|\, y_3(x) \,| &\leq M(L_1 + L_2 p + L_3 q)(L_1 + L_2 p^2 + L_3 q^2)\frac{x^3}{3!} \\
&\leq M(L_1 + L_2 + L_3)^2\frac{x^3}{3!}.
\end{aligned}$$

In general

$$\begin{aligned}
|\, y_n(x) \,| &\leq M\prod_{j=1}^{n-1}(L_1 + L_2 p^j + L_3 q^j)\frac{x^n}{n!} \\
&\leq M(L_1 + L_2 + L_3)^{n-1}\frac{x^n}{n!}, \quad n = 1, 2, 3 \cdots.
\end{aligned}$$

Taking summation over n, we get

$$\begin{aligned}
\left|\sum_{n=0}^{\infty} y_n\right| &\leq \frac{M}{(L_1 + L_2 + L_3)} e^{(L_1+L_2+L_3)x} + \left(1 - \frac{M}{(L_1 + L_2 + L_3)}\right), \quad x \in [0, b], \\
&\leq \frac{M}{(L_1 + L_2 + L_3)} e^{(L_1+L_2+L_3)b} + \left(1 - \frac{M}{(L_1 + L_2 + L_3)}\right).
\end{aligned}$$

*By using Theorem **4.8**, we can conclude that the DJM series solution of eq.* (4.11) *converges uniformly in the interval [0,b].*

4.3 Stability analysis

Now, we consider a particular case of nonlinear eq. (4.11) namely pantograph equation

$$y'(x) = ay(x) + by(px) + cy(qx). \tag{4.12}$$

The following result gives sufficient condition for asymptotic stability of zero solution of eq. (4.12) by using technique of upper bounds.

Theorem 4.14.

If $(a+b+c) < 0$ *then zero solution of eq.* (4.12) *is asymptotically stable.*

Proof

Define

$$z(x) = \max_{0 \leq t \leq x} y^2(x).$$

Now,

$$\begin{aligned} \frac{1}{2}z'(x) &= \frac{1}{2}\frac{d}{dx}(y^2(x)) \\ &= y(x)y'(x) \\ &= y(x)(ay(x) + by(px) + cy(qx)) \\ &= ay^2(x) + by(x)y(px) + cy(x)y(qx) \\ &\leq (a+b+c)z(x) \\ \Rightarrow z(x) &\leq z(0)e^{2(a+b+c)x} \\ \therefore \lim_{t\to\infty} y(x) &= 0, \quad \text{if} \quad (a+b+c) < 0. \end{aligned}$$

This concludes the proof.□

Definition 4.15.

Consider the time-dependent DDE,

$$y'(x) = g(y(x), y(x-\tau_1(x)), y(x-\tau_2(x))), \tag{4.13}$$

where $g : \mathbb{R}\times\mathbb{R}\times\mathbb{R} \to \mathbb{R}$. *The flow* $\phi_x(x_0)$ *is a solution* $y(x)$ *of eq.* (4.13) *with initial condition* $y(x) = x_0,\ x \le 0$. *The point* y^* *is called equilibrium solution of eq.* (4.13) *if* $g(y^*, y^*, y^*) = 0$. ***(a)*** *If, for any* $\epsilon > 0$, *there exist* $\delta > 0$ *such that* $|x_0 - y^*| < \delta \Rightarrow |\phi_x(x_0) - y^*| < \epsilon$, *then the system* (4.13) *is stable (in the Lyapunov sense) at the equilibrium* y^*.***(b)*** *If the system* (4.13) *is stable at* y^* *and moreover,* $\lim_{x\to\infty} |\phi_x(x_0) - y^*| = 0$ *then the system* (4.13) *is said to be asymptotically stable at* y^*.

The following results are analogous to the results in [42].

Theorem 4.16.

Assume that the equilibrium solution y^* *of the equation*

$$y' = g(y(x), y(x-\tau_1^*), y(x-\tau_2^*)), \quad \tau_1^* = \tau_1(x_0), \tau_2^* = \tau_2(x_0)$$

is stable and

$$\|g(y(x), y(x-\tau_1(x)), y(x-\tau_2(x))) - g(y(x), y(x-\tau_1(x_1)), \\ y(x-\tau_2(x_2)))\| < \epsilon_1|x - x_1| + \epsilon_2|x - x_2|,$$

for some $\epsilon_1, \epsilon_2 > 0$ *and* $x, x_1, x_2 \in [x_0, x_0 + c)$, c *is a positive constant, then there exists* $\bar{x} > 0$ *such that the equilibrium solution* y^* *of eq.* (4.13) *is stable on finite time interval* $[x_0, \bar{x})$.

Corollary 4.17.

If the real parts of all roots of $\lambda - a - be^{-\lambda\tau_1^*} - ce^{-\lambda\tau_2^*} = 0$ *are negative, where* $a = \partial_1 f, b = \partial_2 f, c = \partial_3 f$ *evaluated at equilibrium. Then there exist* $\epsilon_c, \bar{x}(> x_0)$, *such that when* $\epsilon_1 < \epsilon_c, \epsilon_2 < \epsilon_c$, the solution $y^* = 0$ of eq. (4.13) is stable on finite time interval $[x_0, \bar{x})$.

4.4 The pantograph equation and its solution

Consider the pantograph equation involving two delays,

$$y'(x) = ay(x) + by(px) + cy(qx), \quad y(0) = 1, \tag{4.14}$$

where $0 < 1$, $0 < q < 1$, $a \in \mathbb{R}$, $b \in \mathbb{R}$ and $c \in \mathbb{R}$.

Integrating eq. (4.14), we get

$$y(x) = 1 + \int_0^x (ay(t) + by(pt) + cy(qt))dt$$

which is of the form eq. (4.5). Applying DJM, we obtain

$$\begin{aligned}
y_0(x) &= 1,\\
y_1(x) &= \int_0^x (ay_0(t) + by_0(pt) + cy_0(qt))dt\\
&= (a+b+c)\frac{x}{1!},\\
y_2(x) &= \int_0^x (ay_1(t) + by_1(pt) + cy_1(qt))dt\\
&= \int_0^x (a(a+b+c)t + b(a+b+c)tp + c(a+b+c)tq)dt\\
&= (a+b+c)(a+bp+cq)\frac{x^2}{2!},\\
y_3(x) &= \int_0^x (ay_2(t) + by_2(pt) + cy_2(qt))dt\\
&= (a+b+c)(a+bq+cq)(a+bp^2+cq^2)\frac{x^3}{3!},\\
&\vdots\\
y_n(x) &= \frac{x^n}{n!}\prod_{j=0}^{n-1}(a+bp^j+cq^j), \quad n = 1,2,3\cdots.
\end{aligned}$$

∴ The DJM solution of eq. (4.14) is

$$\begin{aligned}
y(x) &= y_0(x) + y_1(x) + y_2(x) + \cdots\\
&= 1 + (a+b+c)\frac{x}{1!} + (a+b+c)(a+bp+cq)\frac{x^2}{2!} + \cdots\\
&= 1 + \sum_{n=1}^{\infty}\frac{x^n}{n!}\prod_{j=0}^{n-1}(a+bp^j+cq^j).
\end{aligned} \tag{4.15}$$

From this solution eq. (4.15) of eq. (4.14) we propose a novel special function

$$\Re(a,b,c,p,q,x) = 1 + \sum_{n=1}^{\infty}\frac{x^n}{n!}\prod_{j=0}^{n-1}(a+bp^j+cq^j). \tag{4.16}$$

4.5 Analysis

Theorem 4.18.

If $0<p<1, 0<q<1$, *then the power series*

$$\Re(a,b,c,p,q,x) = 1 + \sum_{n=1}^{\infty} \frac{x^n}{n!} \prod_{j=0}^{n-1} (a + bp^j + cq^j) \tag{4.17}$$

has infinite radius of convergence.

proof.

Suppose

$$a_n = \frac{1}{n!} \prod_{j=0}^{n-1} (a + bp^j + cq^j), \quad n = 1, 2, \cdots.$$

If R is radius of convergence of eq. (4.17) *then by using ratio test* [43]

$$\begin{aligned}
\frac{1}{R} = \lim_{n\to\infty} \left| \frac{a_{n+1}}{a_n} \right| &= \lim_{n\to\infty} \left| \frac{\frac{1}{(n+1)!} \prod_{j=0}^{n} (a + bp^j + cq^j)}{\frac{1}{n!} \prod_{j=0}^{n-1} (a + bp^j + cq^j)} \right| \\
&= \lim_{n\to\infty} \left| \frac{a}{(n+1)} + \frac{bp^n}{(n+1)} + \frac{cq^n}{(n+1)} \right| \\
\Rightarrow \frac{1}{R} &= 0 \quad (\because 0 < p < 1, 0 < q < 1).
\end{aligned}$$

Thus the series has infinite radius of convergence.□

Corollary 4.19.

The power series eq. (4.16) *is absolutely convergent for all x, if* $0 < p < 1, 0 < q < 1$ *and hence it is uniformly convergent on any compact interval on* $\mathbb{R}$.

Proof of the following theorem is trivial.

Theorem 4.20.

For $0 < p < 1, 0 < q < 1$, $a \in \mathbb{R}$, $b \in \mathbb{R}$, $c \in \mathbb{R}$ *and* $m \in \mathbb{N} \cup \{0\}$, *we have*

$$\begin{aligned}
\text{(a)} \quad \frac{d}{dx} \Re(a,b,c,p,q,r^m x) &= r^m (a\Re(a,b,c,p,q,r^m x) + b\Re(a,b,c,p,q,r^m px) \\
&\quad + c\Re(a,b,c,p,q,r^m qx))
\end{aligned}$$

$$\text{and} \quad \text{(b)} \quad \frac{d^m}{dx^m} \Re(a,b,c,p,q,x) = \sum_{n=m}^{\infty} \frac{x^{n-m}}{(n-m)!} \prod_{j=0}^{n-1} (a + bp^j + cq^j).$$

Theorem 4.21.

(Addition Theorem) For $0<p<1, 0<q<1$, $a\in\mathbb{R}$, $b\in\mathbb{R}$, $c\in\mathbb{R}$ *and* $r\in\mathbb{N}\cup\{0\}$, *we have*

$$\Re(a,b,c,p,q,x+y)=\sum_{r=0}^{\infty}\frac{x^r}{r!}\Re^{(r)}(a,b,c,p,q,y)$$

proof.

We have

$$\begin{aligned}\Re(a,b,c,p,q,x+y) &= 1+\sum_{n=1}^{\infty}\frac{(x+y)^n}{n!}\prod_{j=0}^{n-1}(a+bp^j+cq^j)\\ &= 1+\sum_{n=1}^{\infty}\sum_{r=0}^{n}\frac{x^r}{r!}\frac{y^{n-r}}{(n-r)!}\prod_{j=0}^{n-1}(a+bp^j+cq^j).\end{aligned}$$

Define

$$\prod_{j=0}^{n-1}(a+bp^j+cq^j)=1 \quad \textit{for} \quad n=0.$$

to

$$\begin{aligned}\therefore\Re(a,b,c,p,q,x+y) &= \sum_{r=0}^{\infty}\sum_{n=r}^{\infty}\frac{x^r}{r!}\frac{y^{n-r}}{(n-r)!}\prod_{j=0}^{n-1}(a+bp^j+cq^j)\\ &= \sum_{r=0}^{\infty}\frac{x^r}{r!}\sum_{n=r}^{\infty}\frac{y^{n-r}}{(n-r)!}\prod_{j=0}^{n-1}(a+bp^j+cq^j).\end{aligned}$$

to

Using Theorem 20(b), we have

$$\Re(a,b,c,p,q,x+y)=\sum_{r=0}^{\infty}\frac{x^r}{r!}\Re^{(r)}(a,b,c,p,q,y).$$

□

Theorem 4.22.

For $0<p<1, 0<q<1$, $a\geq 0$, $b\geq 0$ *and* $c\geq 0$ *the function* $\Re(a,b,c,p,q,x)$ *satisfies the following inequality*

$$e^{ax}\leq\Re(a,b,c,p,q,x)\leq e^{(a+b+c)x}, \quad 0\leq x<\infty.$$

proof.

Since $0<p<1, 0<q<1$, $a\geq 0$, $b\geq 0$ *and* $c\geq 0$, *we have*

$$\begin{aligned}\prod_{j=0}^{n-1}(a+bp^j+cq^j) &\leq (a+b+c)^n \\ \Rightarrow \frac{x^n}{n!}\prod_{j=0}^{n-1}(a+bp^j+cq^j) &\leq \frac{x^n(a+b+c)^n}{n!}.\end{aligned} \tag{4.18}$$

Taking summation over n, we get

$$\Re(a,b,c,p,q,x) \leq e^{(a+b+c)x}, \quad 0\leq x<\infty.$$

Similarly, we have

$$\begin{aligned}a^n &\leq \prod_{j=0}^{n-1}(a+bp^j+cq^j) \\ \Rightarrow e^{ax} &\leq \Re(a,b,c,p,q,x), \quad 0\leq x<\infty.\end{aligned} \tag{4.19}$$

From eqs (4.18) *and* (4.19), *we get*

$$e^{ax}\leq \Re(a,b,c,p,q,x)\leq e^{(a+b+c)x}, \quad 0\leq x<\infty.$$

Result is illustrated in Figure 4.1. for the values $a=2, b=3, c=4, p=\frac{1}{2}$ and $q=\frac{1}{3}$.

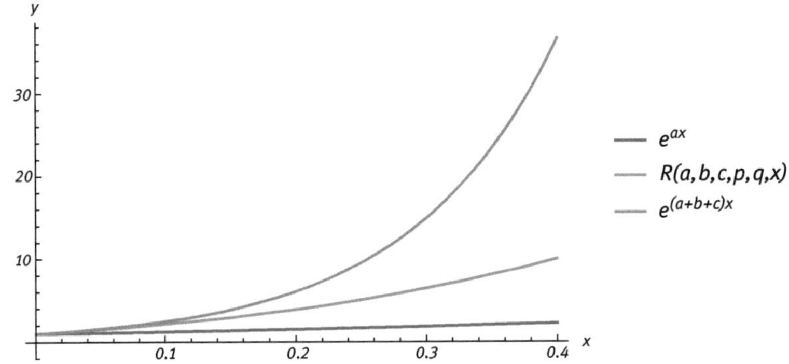

Figure 4.1: Bounds on R(a,b,c,p,q,x) for the values a=2,b=3,c=4,p=1/2a=2,b=3,c=4,p=1/2 and q=1/3.q=1/3.

Theorem 4.23.

$$\int_0^\infty e^{-t}\Re(a,b,c,p,q,e^{-t}t)dt = 1+\sum_{n=1}^\infty \frac{1}{(1+n)^n}\prod_{j=0}^{n-1}(a+bp^j+cq^j).$$

proof.

Consider

$$\begin{aligned}\int_0^\infty e^{-t}\Re(a,b,c,p,q,e^{-t}t)dt &= \int_0^\infty e^{-t}\left(1+\sum_{n=1}^\infty \frac{t^n}{n!}\prod_{j=0}^{n-1}(a+bp^j+cq^j)\right)dt\\ &= \int_0^\infty e^{-t}dt+\sum_{n=1}^\infty \frac{1}{n!}\int_0^\infty e^{-t}t^n dt\prod_{j=0}^{n-1}(a+bp^j+cq^j)\\ &= 1+\sum_{n=1}^\infty \frac{1}{n!}\frac{\Gamma(n+1)}{(n+1)^n}\prod_{j=0}^{n-1}(a+bp^j+cq^j)\\ &= 1+\sum_{n=1}^\infty \frac{1}{(1+n)^n}\prod_{j=0}^{n-1}(a+bp^j+cq^j).\end{aligned}$$

□

Theorem 4.24.

$$\int_x^\infty e^{-t}\Re(a,b,c,p,q,t)dt = \Gamma(1,x)\left(1+\sum_{n=1}^\infty\sum_{k=0}^n \frac{x^k}{k!}\prod_{j=0}^{n-1}(a+bp^j+cq^j)\right).$$

proof.

Consider

$$\begin{aligned}\int_x^\infty e^{-t}\Re(a,b,c,p,q,t)dt &= \int_x^\infty e^{-t}+\sum_{n=1}^\infty\prod_{j=0}^{n-1}\frac{(a+bp^j+cq^j)}{n!}\int_x^\infty e^{-t}t^n dt\\ &= e^{-x}+\sum_{n=1}^\infty\sum_{k=0}^n\prod_{j=0}^{n-1}\frac{(a+bp^j+cq^j)}{n!}\Gamma(n+1,x)\\ &= e^{-x}+\sum_{n=1}^\infty\prod_{j=0}^{n-1}\frac{(a+bp^j+cq^j)}{n!}n!e^{-x}\sum_{k=0}^n\frac{x^k}{k!}\\ &= \Gamma(1,x)\left(1+\sum_{n=1}^\infty\sum_{k=0}^n \frac{x^k}{k!}\prod_{j=0}^{n-1}(a+bp^j+cq^j)\right).\end{aligned}$$

□

Theorem 4.25.

$$\int_0^x e^{-t}\Re(a,b,c,p,q,t)dt = 1+\sum_{n=1}^{\infty}\prod_{j=0}^{n-1}(a+bp^j+cq^j) -\Gamma(1,x)\left(1+\sum_{n=1}^{\infty}\sum_{k=0}^{n}\frac{x^k}{k!}\prod_{j=0}^{n-1}(a+bp^j+cq^j)\right).$$

proof.

Consider

$$\begin{aligned}\int_0^x e^{-t}\Re(a,b,c,p,q,t)dt &= \int_0^x e^{-t}dt+\sum_{n=1}^{\infty}\prod_{j=0}^{n-1}\frac{(a+bp^j+cq^j)}{n!}\int_0^x e^{-t}t^n dt\\ &= 1-e^{-x}+\sum_{n=1}^{\infty}\prod_{j=0}^{n-1}\frac{(a+bp^j+cq^j)}{n!}\gamma(n+1,x)\\ &= 1-e^{-x}+\sum_{n=1}^{\infty}\prod_{j=0}^{n-1}\frac{(a+bp^j+cq^j)}{n!}n!\left(1-e^{-x}\sum_{k=0}^{n}\frac{x^k}{k!}\right)\\ &= 1+\sum_{n=1}^{\infty}\prod_{j=0}^{n-1}(a+bp^j+cq^j)\\ &\quad -\Gamma(1,x)\left(1+\sum_{n=1}^{\infty}\sum_{k=0}^{n}\frac{x^k}{k!}\prod_{j=0}^{n-1}(a+bp^j+cq^j)\right).\end{aligned}$$

4.5.1 The relation between $\Re(a,b,c,p,q,x)$ and Kummer's confluent hypergeometric function

Theorem 4.26.

$$\int_0^x e^{-t}\Re(a,b,c,p,q,t)dt = 1+\Gamma(1,x)\left(\sum_{n=1}^{\infty}\sum_{m=0}^{\infty}\frac{x^{(n+m)}}{n!(n+1)_m}\prod_{j=0}^{n-1}(a+bp^j+cq^j) -\Re(a,b,c,p,q,x)\right)$$

proof.

Consider

$$\begin{aligned}
\int_0^x e^{-t}\Re(a,b,c,p,q,t)dt &= \int_0^x e^{-t} + \sum_{n=1}^{\infty} \frac{\int_0^x e^{-t}t^n dt}{n!} \prod_{j=0}^{n-1}(a+bp^j+cq^j) \\
&= 1-e^{-x} + \sum_{n=1}^{\infty} \frac{\gamma(n+1,x)}{n!} \prod_{j=0}^{n-1}(a+bp^j+cq^j) \\
&= 1-e^{-x} + \sum_{n=1}^{\infty} \frac{(n\gamma(n,x)-x^n e^{-x})}{n!} \prod_{j=0}^{n-1}(a+bp^j+cq^j) \\
&= 1-e^{-x} + \sum_{n=1}^{\infty} \frac{(\gamma(n,x))}{(n-1)!} \prod_{j=0}^{n-1}(a+bp^j+cq^j) \\
&\quad -e^{-x}\sum_{n=1}^{\infty}\frac{x^n}{n!}\prod_{j=0}^{n-1}(a+bp^j+cq^j) \\
&= 1+\sum_{n=1}^{\infty} \frac{(n^{-1}x^n e^{-x}{}_1F_1(1;n+1;x))}{(n-1)!} \prod_{j=0}^{n-1}(a+bp^j+cq^j) \\
&\quad -e^{-x}\Re(a,b,c,p,q,x) \\
&= 1+\Gamma(1,x)\left(\sum_{n=1}^{\infty}\frac{x^n}{n!}{}_1F_1(1;n+1;x)\prod_{j=0}^{n-1}(a+bp^j+cq^j)\right. \\
&\quad \left. -\Re(a,b,c,p,q,x)\right) \\
&= 1+\Gamma(1,x)\left(\sum_{n=1}^{\infty}\sum_{m=0}^{\infty}\frac{x^{(n+m)}}{n!(n+1)_m}\prod_{j=0}^{n-1}(a+bp^j+cq^j)\right. \\
&\quad \left. -\Re(a,b,c,p,q,x)\right).
\end{aligned}$$

Using properties of incomplete gamma functions described in [20], we have following Corollaries.□

Corollary 4.27.

$$\begin{aligned}
\int_0^x e^{-t}\Re(a,b,c,p,q,t)dt &= 1+\sum_{n=1}^{\infty}\frac{x^n}{n!}{}_1F_1(n;n+1;-x)\prod_{j=0}^{n-1}(a+bp^j+cq^j) \\
&\quad -\Gamma(1,x)\Re(a,b,c,p,q,x).
\end{aligned}$$

Corollary 4.28.

$$\int_0^x e^{-t}\Re(a,b,c,p,q,t)dt = 1+\sum_{n=1}^{\infty}\sum_{m=0}^{\infty}\frac{(-1)^m(n)_m x^{(n+m)}}{n!m!(n+1)_m}\prod_{j=0}^{n-1}(a+bp^j+cq^j) -\Gamma(1,x)\Re(a,b,c,p,q,x).$$

Theorem 4.29.

$$\begin{aligned}\int_x^{\infty} e^{-t}\Re(a,b,c,p,q,t)dt &= \Gamma(1,x)+\pi x\Gamma(1,x)\sum_{n=1}^{\infty}\frac{x^{n-1}}{(n-1)!\sin((1+n)\pi)}\\ &\left(\frac{{}_1F_1(1;1+n;x)}{\Gamma(1+n)\Gamma(1-n)}-x^{-n}\frac{{}_1F_1(1-n;1-n;x)}{\Gamma(1-n)}\right)\\ &\prod_{j=0}^{n-1}(a+bp^j+cq^j)+e^x(\Re(a,b,c,p,q,x)-1).\end{aligned}$$

proof.

Consider

$$\begin{aligned}\int_x^{\infty} e^{-t}\Re(a,b,c,p,q,t)dt &= \int_x^{\infty} e^{-t}+\sum_{n=1}^{\infty}\frac{\int_x^{\infty} e^{-t}t^n dt}{n!}\prod_{j=0}^{n-1}(a+bp^j+cq^j)\\ &= e^{-x}+\sum_{n=1}^{\infty}\frac{\Gamma(n+1,x)}{n!}\prod_{j=0}^{n-1}(a+bp^j+cq^j)\\ &= \Gamma(1,x)+\sum_{n=1}^{\infty}\frac{(n\Gamma(n,x)+x^n e^x)}{n!}\prod_{j=0}^{n-1}(a+bp^j+cq^j)\\ &= \Gamma(1,x)+\sum_{n=1}^{\infty}\frac{(\Gamma(n,x))}{(n-1)!}\prod_{j=0}^{n-1}(a+bp^j+cq^j)\\ &+e^x\sum_{n=1}^{\infty}\frac{x^n}{n!}\prod_{j=0}^{n-1}(a+bp^j+cq^j)\\ &= \Gamma(1,x)+\sum_{n=1}^{\infty}\frac{(x^n e^{-x}U(1;1+n;x))}{(n-1)!}\prod_{j=0}^{n-1}(a+bp^j+cq^j)\\ &+e^x(\Re(a,b,c,p,q,x)-1)\\ &= \Gamma(1,x)+x\Gamma(1,x)\sum_{n=1}^{\infty}\frac{x^{n-1}}{(n-1)!}U(1;1+n;x)\\ &\prod_{j=0}^{n-1}(a+bp^j+cq^j)+e^x(\Re(a,b,c,p,q,x)-1)\end{aligned}$$

$$
\begin{aligned}
&= \Gamma(1,x) + \pi x\Gamma(1,x)\sum_{n=1}^{\infty}\frac{x^{n-1}}{(n-1)!\sin((1+n)\pi)} \\
&\quad \left(\frac{{}_1F_1(1;1+n;x)}{\Gamma(1+n)\Gamma(1-n)} - x^{-n}\frac{{}_1F_1(1-n;1-n;x)}{\Gamma(1-n)}\right) \\
&\quad \prod_{j=0}^{n-1}(a+bp^j+cq^j) + e^x(\Re(a,b,c,p,q,x)-1).
\end{aligned}
$$

□

Proof of following Corollaries are immediate from the properties of incomplete gamma functions [20].

Corollary 4.30.

$$
\begin{aligned}
\int_x^{\infty} e^{-t}\Re(a,b,c,p,q,t)dt &= \Gamma(1,x)+\Gamma(1,x)\sum_{n=1}^{\infty}\frac{U(1-n;1-n;x)}{(n-1)!}\prod_{j=0}^{n-1}(a+bp^j+cq^j) \\
&\quad +e^x(\Re(a,b,c,p,q,x)-1).
\end{aligned}
$$

Corollary 4.31.

$$
\begin{aligned}
\int_x^{\infty} e^{-t}\Re(a,b,c,p,q,t)dt &= \Gamma(1,x)+\pi\Gamma(1,x)\sum_{n=1}^{\infty}\frac{1}{(n-1)!\sin((1+n)\pi)} \\
&\quad \left(\frac{{}_1F_1(1-n;1-n;x)}{\Gamma(1-n)} - x^{n}\frac{{}_1F_1(1;1+n;x)}{\Gamma(1-n)\Gamma(1+n)}\right) \\
&\quad \prod_{j=0}^{n-1}(a+bp^j+cq^j) + e^x(\Re(a,b,c,p,q,x)-1).
\end{aligned}
$$

Corollary 4.32.

Using properties of incomplete gamma functions and generalized Laguerre polynomial, we have following expression for $\int_x^{\infty} e^{-t}\Re(a,b,c,p,q,t)dt$:

$$
\begin{aligned}
(i)\quad & \gamma(1,x)+x\Gamma(1,x)\sum_{n=1}^{\infty}\sum_{m=0}^{\infty}\prod_{j=0}^{n-1}(a+bp^j+cq^j)\frac{x^{n-1}}{(n-1)!}\frac{L_m^{(n)}(x)}{(m+1)!} \\
& +e^x(\Re(a,b,c,p,q,x)-1)
\end{aligned}
$$

$$(ii)\qquad \gamma(1,x)+x\Gamma(1,x)\sum_{n=1}^{\infty}\sum_{m=0}^{\infty}\sum_{k=0}^{m}\prod_{j=0}^{n-1}(-1)^k\binom{m+n}{m-k}$$
$$(a+bp^j+cq^j)\frac{x^{n+k-1}}{k!(n-1)!}+e^x(\Re(a,b,c,p,q,x)-1)$$
$$(iii)\qquad \gamma(1,x)+x\Gamma(1,x)\sum_{n=1}^{\infty}\sum_{m=0}^{\infty}\prod_{j=0}^{n-1}(-1)^k\binom{m+n}{m}(a+bp^j+cq^j)$$
$${}_1F_1(-m;n+1;x)\frac{x^{n-1}}{(n-1)!}+e^x(\Re(a,b,c,p,q,x)-1),$$

where $L_m^{(n)}$ is generalized Laguerre polynomial.

Theorem 4.33.

$$\int_0^x e^{-t}\Re(a,b,c,p,q,\lambda t)dt=\gamma(1,x)+\lambda\sum_{n=1}^{\infty}\sum_{m=0}^{n+1}\prod_{j=0}^{n-1}(a+bp^j+cq^j)\frac{\lambda^n}{n!}\frac{\gamma(n+m+1,x)}{m!}(1-\lambda)^m.$$

proof.

We have

$$\begin{aligned}\int_0^x e^{-t}\Re(a,b,c,p,q,\lambda t)dt &= 1-e^{-x}+\sum_{n=1}^{\infty}\prod_{j=0}^{n-1}\frac{(a+bp^j+cq^j)}{n!}\int_0^x e^{-t}(\lambda t)^n dt\\ &= \gamma(1,x)+\sum_{n=1}^{\infty}\prod_{j=0}^{n-1}\frac{(a+bp^j+cq^j)}{n!}\gamma(n+1,\lambda x).\end{aligned}$$

By using the property Gautschi et al. [44] $\gamma(n,\lambda x)=\lambda^n\sum_{m=0}^{n}\frac{\gamma(n+m,x)}{m!}(1-\lambda)^m$, *we get*

$$\begin{aligned}\int_0^x e^{-t}\Re(a,b,c,p,q,\lambda t)dt &= \gamma(1,x)+\lambda\sum_{n=1}^{\infty}\sum_{m=0}^{n+1}\prod_{j=0}^{n-1}(a+bp^j+cq^j)\\ &\quad \frac{\lambda^n}{n!}\frac{\gamma(n+m+1,x)}{m!}(1-\lambda)^m.\end{aligned}$$

□

Theorem 4.34.

$$\int_x^{\infty} e^{-t}\Re(a,b,c,p,q,\lambda t)dt = \Gamma(1,x)+\lambda\sum_{n=1}^{\infty}\sum_{m=0}^{n+1}\prod_{j=0}^{n-1}(a+bp^j+cq^j)\frac{\lambda^n}{n!}\frac{\Gamma(n+m+1,x)}{m!}(1-\lambda)^m.$$

proof.

$$\begin{aligned}\int_x^{\infty} e^{-t}\Re(a,b,c,p,q,\lambda t)dt &= e^{-x}+\sum_{n=1}^{\infty}\prod_{j=0}^{n-1}\frac{(a+bp^j+cq^j)}{n!}\int_x^{\infty} e^{-t}(\lambda t)^n dt\\ &= \Gamma(1,x)+\sum_{n=1}^{\infty}\prod_{j=0}^{n-1}\frac{(a+bp^j+cq^j)}{n!}\Gamma(n+1,\lambda x).\end{aligned}$$

By using the property Gautschi et al. [44]$\Gamma(n,\lambda x)=\lambda^n\sum_{m=0}^{n}\frac{\Gamma(n+m,x)}{m!}(1-\lambda)^m$, *we obtain*

$$\int_x^{\infty} e^{-t}\Re(a,b,c,p,q,\lambda t)dt=\Gamma(1,x)+\lambda\sum_{n=1}^{\infty}\sum_{m=0}^{n+1}\prod_{j=0}^{n-1}(a+bp^j+cq^j)\frac{\lambda^n}{n!}\frac{\Gamma(n+m+1,x)}{m!}(1-\lambda)^m.$$

□

4.6 Generalizations to fractional-order DDE

Consider the fractional DDE with proportional delay

$$D_0^{\alpha}y(x)=ay(x)+by(px)+cy(qx),\quad y(0)=1, \tag{4.20}$$

where $0<\alpha\leq 1$, $0<p<1, 0<q<1$, $a\in\mathbb{R}$, $b\in\mathbb{R}$ and $c\in\mathbb{R}$. Equivalently

$$y(x)=1+I^{\alpha}(ay(x)+by(px)+cy(qx)).$$

The DJM solution of eq. (4.20) is

$$y(x)=1+\sum_{n=1}^{\infty}\frac{x^{\alpha n}}{\Gamma(\alpha n+1)}\prod_{j=0}^{n-1}(a+bp^{\alpha j}+cq^{\alpha j}). \tag{4.21}$$

We denote the series in eq. (4.21) by

$$\Re_\alpha(a,b,c,p,qx) = 1 + \sum_{n=1}^{\infty} \frac{x^{\alpha n}}{\Gamma(\alpha n+1)} \prod_{j=0}^{n-1} (a + bp^{\alpha j} + cq^{\alpha j}).$$

Theorem 4.35.

If $0 < q < 1$, *then the power series*

$$\Re_\alpha(a,b,c,p,qx) = 1 + \sum_{n=1}^{\infty} \frac{x^{\alpha n}}{\Gamma(\alpha n+1)} \prod_{j=0}^{n-1} (a + bp^{\alpha j} + cq^{\alpha j}),$$

is convergent for all finite values of x.

proof.

Result follows immediately by ratio test.□

4.7 Conclusions

In this paper, we have obtained a new special function arising from pantograph equation with two delays. The solution is obtained by applying the iterative scheme namely DJM. The existence, uniqueness, stability and convergence results for the time-dependent DDE are presented in this paper. The new special function exhibits different properties and relations with other functions. The generalization to fractional-order case is also presented. We hope that the researchers will get motivated from this work and work on more properties of this new special function.

Funding: S. Bhalekar acknowledges CSIR, New Delhi for funding through Research Project [25(0245)/15/EMR-II].

References

[1] Hale JK, Lunel SM. Introduction to functional differential equations Vol. 99. New York: Springer Science & Business Media, 2013.
[2] Smith H. An introduction to delay differential equations with applications to the life sciences Vol. 57. New York: Springer Science & Business Media, 2010.
[3] Erneux T. Applied delay differential equations Vol. 3. New York: Springer Science & Business Media, 2009.
[4] Lakshmanan M. Senthilkumar DV. Dynamics of nonlinear time-delay systems. Berlin: Springer Science & Business Media, 2011.
[5] https://www.youtube.com/watch?v=d4Zic91CmRU\&feature=player_embedded/.
[6] http://www.railsystem.net/pantograph/.
[7] Ockendon J, Tayler AB. The dynamics of a current collection system for an electric locomotive. Proc R Soc Lond A Math Phys Eng Sci. 1971;322:447–468.

[8] Kato T, McLeod JB. The functional-differential equation y(x) = ay(x) + by(x). Bull Am Math Soc. 1971;77:891–937.
[9] Liu Y. Numerical investigation of the pantograph equation. Appl Numer Math. 1997;24(2):309–317.
[10] Bhalekar S, Patade J. 2016 Novel special function obtained from a delay differential equation. arXiv preprint arXiv:1608.03959.
[11] Iserles A. On the generalized pantograph functional-differential equation. Eur J Appl Math. 1993;4(1):1–38.
[12] Bellen A, Guglielmi N, Torelli L. Asymptotic stability properties of θ-methods for the pantograph equation. Appl Numer Math. 1997;24(2):279–293.
[13] Derfel G, Iserles A. The pantograph equation in the complex plane. Journal of Mathematical Analysis and Applications. 1997;213(1):117–132.
[14] Koto T. Stability of RungeKutta methods for the generalized pantograph equation, Numer. Math. 84(1999):23347.
[15] Liu M. Z., Li D. Properties of analytic solution and numerical solution of multi-pantograph equation. Applied Mathematics and Computation. 155(3):853–871 2004.
[16] Dehghan M., Shakeri F. The use of the decomposition procedure of Adomian for solving a delay differential equation arising in electrodynamics. Physica Scripta. 78(6):065004 2008.
[17] Saadatmandi A., Dehghan M. Variational iteration method for solving a generalized pantograph equation. Computers & Mathematics with Applications. 58(11):2190–2196 2009.
[18] Li D., Liu M. Z. RungeKutta methods for the multi-pantograph delay equation. Applied mathematics and computation. 2005;163(1):383–395.
[19] Sezer M, Sahin N. Approximate solution of multi-pantograph equation with variable coefficients. J Comput Appl Math. 2008;214(2):406–416.
[20] Magnus W, Oberhettinger F, Soni R. Formulas and theorems for the special functions of mathematical physics Vol. 52. Springer Science & Business Media, 2013.
[21] Kilbas AA, Srivastava HM, Trujillo JJ. Theory and applications of fractional differential equations. Amsterdam: Elsevier, 2006.
[22] Kac V, Cheung P. Quantum calculus. New York: Springer, 2002.
[23] Rudin W. Principles of mathematical analysis. New York: McGraw-Hill, 1964.
[24] Daftardar-Gejji V, Jafari H. An iterative method for solving non linear functional equations. J Math Anal Appl. 2006;316:753–763.
[25] Daftardar-Gejji V, Bhalekar S. Solving fractional diffusion-wave equations using the new iterative method. Frac Calc Appl Anal. 2008;11:193–202.
[26] Bhalekar S, Daftardar-Gejji V. New iterative method: Application to partial differential equations. Appl Math Comput. 2008;203:778–783.
[27] Daftardar-Gejji V, Bhalekar S. Solving fractional boundary value problems with Dirichlet boundary conditions. Comput Math Appl. 2010;59:1801–1809.
[28] Mohyud-Din ST, Yildirim A, Hosseini MM. An iterative algorithm for fifth-order boundary value problems. World Appl Sci J. 2010;8:531–5.
[29] Bhalekar S, Daftardar-Gejji V. Solving evolution equations using a new iterative method. Numer Methods Part Differ Equ. 2010;26:906–916.
[30] Bhalekar S, Daftardar-Gejji V. Solving a system of nonlinear functional equations using revised new iterative method. Int J Comput Math Sci. 2012;6:127–131.
[31] Patade J, Bhalekar S. Approximate analytical solutions of Newell-Whitehead-Segel equation using a new iterative method. World J Modell Simul. 2015;11(2):94–103.
[32] Bhalekar S, Patade J. An analytical solution of fishers equation using decomposition method. Am J Comput Appl Math. 2016;6(3):123–127.

[33] Bhalekar S, Daftardar-Gejji V. Int J Differ Equ. Article ID 975829 2012 Solving fractional order logistic equation using a new iterative method 2010.
[34] Bhalekar S, Daftardar-Gejji V. Numeric-analytic solutions of dynamical systems using a new iterative method. J Appl Nonlinear Dyn. 2012;1:141–158.
[35] Daftardar-Gejji V, Sukale Y, Bhalekar S. A new predictor-corrector method for fractional differential equations. Appl Math Comput. 2014;244:158–182.
[36] Daftardar-Gejji V, Sukale Y, Bhalekar S. Solving fractional delay differential equations: a new approach. Fract Calc Appl Anal. 2015;16:400–418.
[37] Patade J, Bhalekar S. A new numerical method based on Daftardar-Gejji and Jafari technique for solving differential equations. World J Modell Simul. 2015;11(4):256–271.
[38] Bhalekar S, Patade J. A novel third order numerical method for solving Volterra integro-differential equations. arXiv preprint:1604.08863v1 [math.NA].
[39] Bhalekar S, Daftardar-Gejji V. Convergence of the new iterative method. Int J Differ Equ 2011;2011:1–10.
[40] Bellen A, Zennaro M. Numerical methods for delay differential equations. Oxford: University Press, 2013.
[41] Bhalekar S, Patade J. Analytical solutions of nonlinear equations with proportional delays. Appl Comput Math. 2016;15(3):331–445.
[42] Deng W, Wu Y, Li C. Stability analysis of differential equations with time-dependent delay. Int J Bifurcation Chaos. 2006;16(02):465–472.
[43] Apostal TM. Mathematical analysis. Reading: Addison-Wesley, 1964.
[44] Gautschi W, Frank EH, Nico MT. Expansions of the exponential integral in incomplete gamma functions. Appl Math Lett. 2003;16(7):1095–9.

M. Alcolea Palafox

5 Computational chemistry applied to vibrational spectroscopy: A tool for characterization of nucleic acid bases and some of their 5-substituted derivatives

Abstract: Computational chemistry can be applied to vibrational spectroscopy in different ways, such as for a better characterization and assignment of all the bands of the experimental spectra, as a tool in the identification of the tautomers present in the gas phase and in the solid state through their spectra and for the simulation of the solid and liquid phase of a compound and the consequent simulation and interpretation of their spectra. In the present study, as an example of the applicability of computational chemistry, the structure and spectra of cytosine and uracil nucleic acid bases and two cytosine derivatives are shown. The FTIR and Raman spectra were analysed with the support of ab initio (Hartree-Fock (HF), MP2) and density functional theory (DFT) (B3LYP, PBE, B-P, etc.) calculations using several basis sets and several scaling equations. The calculations predict an easier tautomerization of cytosine than uracil molecule, but the tautomerization is hindered in the 5-bromocytosine molecule. Thus, in the solid state, this molecule only exists in the amino-oxo tautomeric form.

Keywords: applications of computational chemistry, cytosine molecule, scaling

5.1 Introduction

The motivation for the application of computational chemistry to vibrational spectroscopy is to make it a more practical tool [1]. From a practical point of view, the main disadvantage of vibrational spectroscopy is the lack of a direct spectrum–structure relation. This makes it impossible or difficult to determinate the structure of a molecule from its vibrational spectrum. However, vibrational spectroscopy has a number of advantages over other spectroscopic methods, such as Nuclear Magnetic Resonance (NMR). Most of these relate to the inherently greater sensitivity of vibrational spectroscopy, which makes it possible to detect very small amounts. Further advantages are the wider scope of vibrational spectroscopy, e.g. its applicability to solids, liquids and gasses, as well as to adsorbed layers, etc. Instrument costs of IR spectroscopy are also generally lower than for other spectroscopic techniques. It is thus clear that many of the advantages of vibrational spectroscopy could be increased if a method could be found to reliably predict vibrational spectra. Such a

https://doi.org/10.1515/9783110467215-005

method could be used to calculate the expected spectra of proposed structures. Comparison with the observed spectra would confirm the identity of a product, even that of a completely new molecule. Density functional theory (DFT) quantum chemical methods are the most suitable for this purpose, and for many tasks the B3LYP/DFT method is the most commonly used today [2–6].

In general, the computation of the vibrational spectrum of a polyatomic molecule of even modest size is lengthy. The most accurate of the quantum chemical methods is still too expensive and cumbersome to apply in routine research. Thus, one may be forced to work at low level, and consequently, one must expect a high overestimation of the calculated vibrational frequencies. This overestimation can be significantly reduced by the use of transferable empirical parameters for the calculated wavenumbers [1]. The scaling is therefore designed to correct the calculated harmonic wavenumbers to be compared with the anharmonic wavenumbers found by experiment. Thus, the first step after the computation of vibrational wavenumbers is to correct the systematic errors in their values, i.e. the scaling. The present manuscript shows the use of this scaling in the spectra of the cytosine and uracil molecules, which can permit an accurate assignment of wavenumbers.

5.2 Molecules under study

A special characteristic of the molecules under study is their tautomerism. Nucleic acid bases are constituents of DNA and RNA, and they play important roles in the transcription of the genetic code. Although for this process, they naturally occur as one predominant isomer, other minor tautomeric forms also exist. In the famous 1953 publication [7], Watson and Crick stated the importance of tautomeric forms of pyrimidine and purine nucleic acid bases with respect to three-dimensional stacking in DNA. This phenomenon and the hypothesis that rare tautomers may be responsible for DNA mutations have encouraged many chemists to carry out theoretical and experimental studies on the structure of nucleobases in different environments. For this purpose, computational methods have been extensively used as important tools for the interpretation of vibrational spectra [8, 9]. Another interest of these molecules is because the experimental IR wavenumbers in the gas phase are available for the uracil molecule and in an Ar matrix for cytosine molecule. This feature enables us to calculate accurate scaling factors/equations to be used in their derivatives and in related molecules, which facilitates a good match between the scaled and experimental wavenumbers.

Cytosine (in short Cy, Figure 5.1**b**), also named as 4-amino-2(1H)-pyrimidinone or 4-amino-2-hydroxypyrimidine, is a pyrimidine base and a constituent of nucleotides and, as such, one member of the base pair guanine–cytosine. Uracil (in short U, Figure 5.1**a**) is also a pyrimidine base and a constituent of nucleotides and, as such,

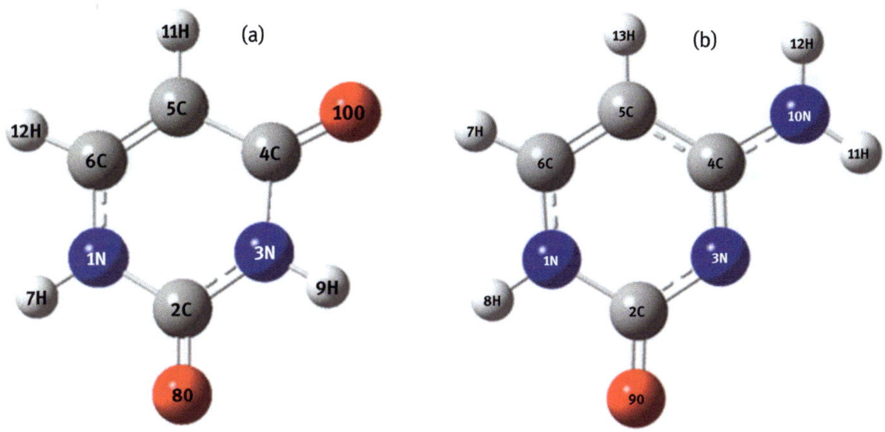

Figure 5.1: Structures of (a) uracil and (b) cytosine with the labels of their atoms.

one member of the base pair adenine–uracil. They belong to the group of the most important pyrimidines that play a fundamental role in the structure and function of enzymes and drugs. The importance of Cy and their derivatives is currently indicated by the considerable number of publications devoted to them appearing in the bibliography [10–14].

Cy contains two labile protons and five conjugated tautomeric sites. Thus, it can exist in various tautomeric forms differing from each other by the position of the proton, which may be bound to either the ring nitrogen atoms or the oxygen atom: amino-oxo (C1), amino-hydroxy (C2a, C2b), imino (C3a, C3b) and (C4) 3H-oxo forms. Of all the possible combinations [15], the six tautomers of Figure 5.2 are the most important and studied tautomers. This molecule is the only nucleobase where the *enol* tautomer C2b is more stable than the *keto* C1 tautomer in the isolated state. There is a great interest in the tautomerism of Cy because of this property [15, 16].

The bioactivity of 5-substituted Cy derivatives also generates exceptional interest in their biochemistry and pharmacology, and they are the most interesting and widely studied Cy derivatives. Among these compounds, 5-halogenated derivatives are of special relevance. Transformation of Cy into 5-halogen-cytosine significantly changes its chemical and spectroscopic properties, as well as its in vivo activity. The halogenated pyrimidines were synthesized in the 1950s as potential antitumor agents. Chlorinated pyrimidines are effective mutagens, clastogens and toxicants, as well as extremely effective inducers of sister-chromatid exchanges [17]. These chlorinated adducts can be mutagenic or perturb DNA–protein interactions [18]. Considering the importance of 5-halogencytosine derivatives for medicinal chemistry, their vibrational spectra, taken for low-temperature matrices and for the polycrystalline state, have not been much explored. Although the

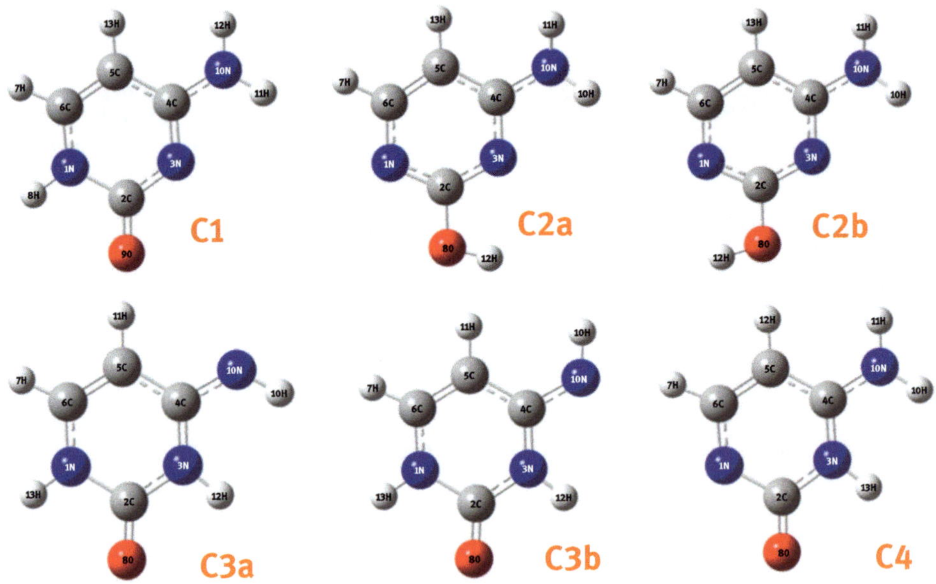

Figure 5.2: Cytosine tautomers with standard numbering and adopted nomenclature: Nonaromatic 2-oxo form (C1), aromatic 2-hydroxy *trans* form (C2a), aromatic 2-hydroxy *cis* form (C2b), non-aromatic 4-imino *cis* form (C3a), non-aromatic 4-imino *trans* form (C3b) and non-aromatic 3H-amino-oxo form (C4).

vibrational spectra of 5-fluoro, 5-chloro and 5-bromo-cytosines have been reported previously on the basis of normal coordinate analysis on semiempirical [19] and DFT [20] methods, there is much controversy in their assignments. Thus, 5-chlorocytosine (5-ClCy) and 5-bromocytosine (5-BrCy) are briefly analysed in the present manuscript.

5.3 Computational methods

Quantum chemical methods are commonly used to analyse and to interpret the molecular structure and the vibrational spectra of compounds. Its use is shown in the present study of the uracil and cytosine nucleic acid bases, and in two of their derivatives. The molecular structure of Cy is analysed from the data available in the bibliography that have been determined theoretically by quantum chemical methods and experimentally by X-ray diffraction. The values of the geometrical parameters are collected in Table 5.1. They are the most accurate calculated today for this compound.

Among the quantum chemical methods, DFT [25, 26] results were selected as the most appropriate. DFT methods provide a very good overall description of medium-size molecules. Moreover, for the wavenumber calculations [1, 27], they appear more accurate than Hartree-Fock (HF) and MP2 methods, and at lower computational cost. Among the DFT methods, the Becke's three-parameter exchange functional (B3) [28]

Table 5.1: Selected equilibrium geometries of cytosine, bond lengths in Å and bond angles in degrees. The calculated values are with different ab initio and DFT methods and basis sets.

Parameters	HF	MP2		B3LYP	B-P	BH-LYP	CCSD[c]	Exp.[d]
	6-311+G	6-311++G	cc-pVTZ[b]	6-311++G	6-311++G	6-311++G	6-31G(d,p)	
	(2d,p)	(2df,2pd)[a]		(3df,pd)	(2df,2pd)[a]	(2df,2pd)[a]		
Bond lengths								
N1-C2	1.3974	1.413	1.4083	1.4226	1.431	1.404	1.413	1.399
C2-N3	1.3604	1.373	1.3684	1.3664	1.373	1.356	1.385	1.356
N3=C4	1.2949	1.313	1.3079	1.3142	1.324	1.300	1.313	1.334
C4-C5	1.4430		1.4259	1.4357			1.457	1.426
C5=C6	1.3387		1.3503	1.3528			1.354	1.337
N1-C6	1.3449		1.3461	1.3495			1.363	1.364
C2=O	1.1938	1.218	1.2147	1.2139	1.226	1.201	1.220	1.237
C4-N9	1.3480	1.362	1.3556	1.3562	1.370	1.342	1.368	1.334
Bond angles								
N-C2-N	116.75		115.98	116.05			116.3	
C2-N=C4	120.35		120.21	120.66			119.6	
N3=C4-C5	123.84		124.25	123.75			124.6	
C2-N1-C6	123.02		123.75	123.28			123.7	
N3-C2=O	124.89		125.18	125.61			124.8	
Torsional angles								
N3=C4-N9-H12	10.08		11.7	6.96			13.7	
C5-C4-N9-H13			−18.9	−10.91			−25.4	
C2-N3=C4-N9	178.26		176.6	178.71			176.9	

[a] Ref [21]. [b] Ref [22]. [c] From ref [23]. [d] From X-ray and neutron diffraction data summarized in a statistical survey of the Cambridge Structural Database [23, 24].

in combination with the correlation functional of Lee, Yang and Parr (LYP) [29], i.e. B3LYP, appears as best and most frequently used today. The results obtained with several basis sets, differing in size and contraction, are shown in the Tables 5.1–5.5 of the manuscript. The 6-311++G(3df,pd) basis set is in general too large for wavenumber calculations, due to the computational memory required. The 6-31G(d,p) and 6-311+G(2d,p) basis sets are optimum for this purpose. All the results were determined with the GAUSSIAN 09 [30] program package.

5.4 Scaling

The vibrational wavenumbers are usually calculated using the simple harmonic oscillator model. Therefore, they are typically larger than the fundamentals observed experimentally. This overestimation in the wavenumbers also depends on the type of vibrational mode and the range considered.

In general, Figure 5.3 shows the error in the wavenumber of a vibrational mode as computed in a variety of molecular environments, with different methods and sized basis sets. The vertical axis shows the difference between the true value in a given molecule of some particular vibrational mode, and the value computed with different methods and various sizes of self-consistent field (SCF) basis sets. The errors in computing the chosen vibrational mode in many different molecules are found to fall within the shaded area of the diagram. For small basis sets (or semiempirical calculations), the error, or range of uncertainty about the "true" value obtained from experiment, is largely increased. The convergence limit, approached by very large

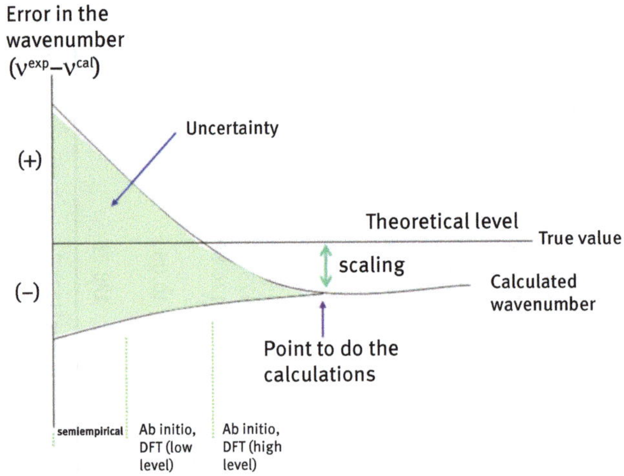

Figure 5.3: Schematic representation of the error in calculating a vibrational mode in a variety of molecular environments. For wide families of systems, the error is expected to fail within the shaded area.

basis sets, still differs from the true value, but this residual error has been found empirically to be remarkably constant for a given parameter and is independent of the molecule studied. The calculations can be done efficiently at the point marked in the figure, and the residual error can be removed by the use of scaling, and therefore give rise to an accurate predicted wavenumber [1].

5.5 Applications of computational chemistry to vibrational spectroscopy

Among all the possibilities, only four applications are shown here, but they are among the most used today. These applications were performed on the uracil, cytosine, 5-chlorocytosine and 5-bromocytosine molecules. These are the following:

a. Characterization of all the normal modes of a molecule.
b. Accurate assignment of all the bands of a spectrum.
c. Identification of the tautomers present in the gas phase and in the solid state of a compound.
d. Simulation of the crystal unit cell of a compound and the interpretation of its vibrational spectra, which remarkably improves the accuracy in the assignment of its spectra in the solid state.

5.5.1 Characterization of all the normal modes of a molecule

Computations at the B3LYP/6-31G(d,p) level has been used here for the characterization of all the normal ring modes in Cy (Figure 5.4) and U (see Ref. [8].). In the Cy molecule, only 30 ring modes are included in the figure. The number of the mode in increasing order appears in the centrum of each ring. Below the plot of each mode appears its calculated wavenumber (in cm^{-1}) and its main characterization. The atomic displacements for each computed wavenumber are determined as XYZ coordinates in the standard orientation, and they are plotted here to identify each vibration.

It is interesting to observe that a few skeletal ring modes retain a certain resemblance to the pseudo-normal modes of a hypothetical C_6 ring in the skeletal modes of benzene, although the uracil ring has no symmetry within the molecular plane. The previous characterization of the normal modes of a molecule by using computational methods led to an appropriate assignment of its bands, which in the present study correspond to the uracil and cytosine molecules and two of their derivatives.

5.5.2 Accurate assignment of all the bands of a spectrum

The computational methods provide information about the harmonic vibrational wavenumbers (in cm^{-1}), absolute infrared (IR) intensities (KM/Mole), Raman

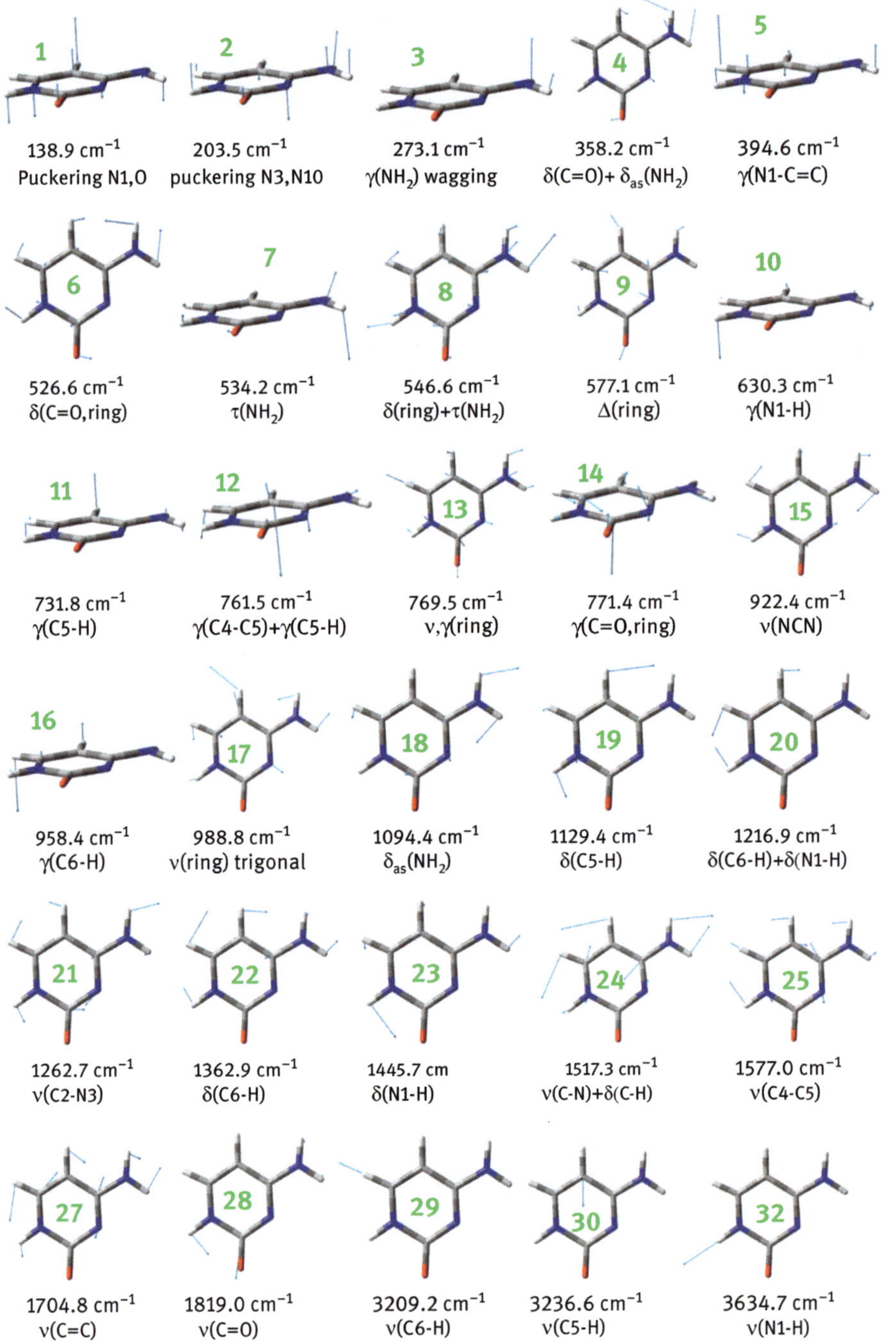

Figure 5.4: Characterization of the normal vibrational modes in the *keto* tautomer C1 of the cytosine molecule at the B3LYP/6-31G(d,p) level.

scattering activities (Å4/amu), Raman depolarization ratios for plane and unpolarized incident light, force constants (mDyne/Å), reduced masses (AMU) and normal coordinates displacement vectors of a molecule after its structure is optimized. All this information helps in the assignment of the bands in the experimental IR and Raman spectra.

5.5.2.1 Scaling procedures

The calculated wavenumbers generally have a large error for many reasons [1, 8, 9], such as anharmonicity, errors in the computed geometry, Fermi resonance, solvent effects, etc., and it is necessary to correct them. For this purpose, the first step after the calculation of the vibrational wavenumbers is the scaling, which helps to carry out an accurate assignment. To improve these computed wavenumbers, two scaling procedures can be used [8, 27].

(a) The first procedure uses a single overall scale factor for the calculated wavenumbers, ν^{exp}/ν^{calc}. This is the easiest way and thus is the procedure generally used in the bibliography to scale wavenumbers. To correct the overestimation of the calculated wavenumbers, several authors have reported scale factors for different levels. The most complete set of values has been determined by Scott and Radom [31], with the particularity of using two scale factors, one for the high- and medium-wavenumber vibrations, and another for low wavenumbers.

(b) A remarkable improvement in the accuracy of the scaled wavenumbers is obtained if a linear relationship is established between the calculated and experimental wavenumbers. This procedure called linear scaling equation (LSE) uses a scaling equation to correct the computed wavenumbers of a molecule at a specific level of theory. This scaling equation, obtained previously for a specific molecule (or group of molecules) with experimental gas-phase wavenumbers available (or Ar matrix values), can be used in related molecules and in their derivatives due to the good transferability of their parameters, which enables accurate assignment of the experimental values for many compounds. This procedure, developed by us, represents a compromise between accuracy and simplicity, and the results obtained are accurate enough for the standard today in large molecules [1, 8, 9].

Following this last procedure, the scaling equations for Cy and U molecules collected in Table 5.2 were obtained at different levels of computation.

5.5.2.2 Assignment of all the bands in the cytosine molecule

The vibrational bands computed in tautomer C1 of the Cy molecule with B3LYP and two different basis sets are examined in Table 5.3. The first column refers to the numbers assigned to the calculated vibrations, and they are given in increasing order of wavenumbers. These normal modes of Cy structure appear plotted in Figure 5.4. The second and the third column list the calculated wavenumbers with the 6-31G(d,p)

Table 5.2: Linear scaling equations $\nu^{scaled} = a + b \cdot \nu^{calculated}$ obtained in cytosine and uracil molecules.

Levels of computation	*a*	*b*	Correlation coeffient, r
Cytosine molecule			
HF/6-31+G(d,p)	−4.1	0.8965	0.9997
HF/6-31+G(2d,p)	−14.3	0.9053	0.9997
B3LYP/6-31+G(d,p)	16.3	0.9560	0.9999
B3LYP/6-31+G(2d,p)	6.2	0.9631	0.9999
B3LYP/6-311+G(2d,p)	4.8	0.9671	0.9999
Uracil molecule			
HF/6-31G(d,p)	5.7	0.8928	0.9997
HF/6-31++G(d,p)	10.5	0.8938	0.9998
MP2/6-31G(d)	34.5	0.9372	0.9996
BP86/6-31G(d,p)	46.0	0.9678	0.9998
BLYP/6-31G(d,p)	46.4	0.9718	0.9998
B3P86/6-31G(d,p)	34.1	0.9389	0.9999
B3LYP/6-31G(d)	30.8	0.9468	0.9999
B3LYP/6-31G(d,p)	34.6	0.9447	0.9999
B3LYP/6-311+G(2d,p)	30.8	0.9538	0.9999
B3LYP/6-311++G(3df,pd)	31.9	0.9512	0.9999
B3LYP/dgdzvp	39.2	0.9472	0.9999
B3PW91/6-31G(d)	30.1	0.9421	0.9999
B3PW91/6-31G(d,p)	34.9	0.9393	0.9999

and 6-311+G(2d,p) basis sets, respectively. The relative IR intensities of the fourth column were obtained by normalizing the computed values to the intensity of the strongest line, no. 28. Small differences appear between the calculated values with both basis sets. The value of the IR intensity helps to carry out the match of theoretical to experimental wavenumbers. Although wavenumber calculations on Cy have been reported with other quantum chemical methods [22, 32], the values shown in Table 5.3 represent the most accurate today.

The characterization established by B3LYP for each calculated wavenumber is shown in the fifth column. The percent contribution of the different modes to a computed wavenumber appears in parentheses. Contributions lower than 10 % were not considered. The sixth and seventh columns collect the experimental wavenumbers reported in neon and argon [33, 34] matrices. The most detailed study corresponds to an Ar matrix [34], and thus their values were selected as reference.

The scaled wavenumbers of the eight to tenth columns were obtained following the scaling procedures mentioned above. These scaled wavenumbers refer to an isolated molecule, and thus they can be directly compared to the experimental wavenumbers in an Ar matrix. From this match, the experimental bands can be assigned. This scaling was carried out in the calculated wavenumbers of tautomers

Table 5.3: Comparison of the calculated harmonic wavenumbers ($\nu^{cal.}$, cm^{-1}) at the B3LYP/6-31+G(d,p) level, the relative (A, %) infrared intensities, and scaled wavenumbers with the experimental values ($\nu^{exp.}$, cm^{-1}) with the % relative IR intensity in parenthesis. The characterization obtained (in parenthesis are the % PED contribution) in the normal ring modes of the cytosine molecule is also included together with the values of the absolute error obtained by different procedures.

No.	Calculated				$\nu^{exp., c}$		Scaled[d]			Absolute error ($\nu^{cal.} - \nu^{exp.}$) in cm^{-1}				
										Calculated[e]		With a factor[f]	With LSE[g]	
	$\nu^{cal.}$	$\nu^{cal., a}$	A	Characterization[b]	Ne	Ar	(a)	(b)	(c)	(d)	(e)	(f)	(g)	(h)
1	134	134	0	(100%) puckering N1			134	144	138					
2	202	202.3	1	(100%) puckering N3			156	165	202					
3	156	257.4	31	(95%) $\gamma(NH_2)$ wagging	220 (47)	236 (30)	202	209	245	−34	12	−34	−27	9
4	358	359.2	0	(43%) δ(C=O) + (42%) $\delta_{as}(NH_2)$ + (15%) δ,Γ(ring)	342 (2)	343 (5)	358	359	351	15	15	15	16	8
5	398	398.9	3	(45%) γ(N1-C=C) + (30%) γ (ring)+ (25%) $\tau(NH_2)$	397 (6)	400 (4)	399	397	393	−2	2	−1	−3	−7
6	527	520.3	2	(80%) $\tau(NH_2)$	511 (56)	507 (22)	528	520	510	20	16	21	13	3
7	533	533.5	0	(60%) δ(C=O)+ (25%) δ,Γ(ring) + (15%) $\delta_{as}(NH_2)$	525 (32)	520 (57)	534	526	519	13	13	14	6	−1
8	545	547.2	0	(80%) δ(ring) mainly in N1, N3 + (20%) $\tau(NH_2)$	531 (17)	535 (16)	546	537	532	10	11	11	2	−3
9	577	580.7	0	(95%) δ(ring) mainly in O8	571 (3)	568 (8)	578	568	564	9	11	10	0	−4
10	624	618.8	8	(70%) γ(N1-H) + (25%) γ(C4- N9)	614 (27)	613 (17)	625	613	608	11	12	12	0	−5
11	724	722.8	5	(62%) γ(C5-H) + (27%) γ(ring) + (11%) $\tau(NH_2)$	711 (3)	710 (3)	696	708	708	14	19	−14	−2	−2
12	763	765.5	1	(75%) δ,γ(ring) + (25%) $\delta_s(NH_2)$	717 (8)	717 (8)	734	746	746	46	51	17	29	29

(continued)

Table 5.3: (*continued*)

No.	Calculated				$\nu^{exp., c}$		Scaled[d]			Absolute error ($\nu^{cal.} - \nu^{exp.}$) in cm^{-1}				
										Calculated[e]		With a factor[f]	With LSE[g]	
	$\nu^{cal.}$	$\nu^{cal., a}$	*A*	Characterization[b]	Ne	Ar	(a)	(b)	(c)	(d)	(e)	(f)	(g)	(h)
13	769	770.4	1	(43%) γ(C4-C5)+ (30%) γ(C5-H) + (27%) γ, δ(ring)	749 (5)	747 (4)	739	751	753	22	28	−8	4	6
14	773	789.2	5	(65%) γ(C-N3=C) mainly in C2 + (35%) γ(ring)	767 (3)	767 82)	743	755	766	6	22	−24	−12	−1
15	924	923.5	1	(65%) ν(ring) + (35%) $\delta_{as}(NH_2)$			888	900	900					
16	962	963.8	0	(75%) γ(C6-H) + (25%) γ(C5-H)			925	936	933					
17	988	989.6	0	(80%) ν(ring) + (20%) $\delta_{as}(NH_2)$	948 (0)	955 (1)	950	961	958	33	33	−5	6	3
18	1088	1094.9	7	(50%) $\delta_{as}(NH_2)$ + (35%) δ(C=O) + (15%) δ(ring)	1085 (10)	1083 (12)	1046	1056	1063	5	14	−37	−27	−20
19	1128	1126.6	0	(50%) δ(C5-H)+(30%) δ(N1-H) +(20%) ν(C=C-N)	1103 (3)	1091 (6)	1084	1095	1093	37	37	−7	4	2
20	1220	1213.3	6	(50%) δ(HC6-NH)+(35%) δ(C5-H)+(15%) ν(N-C6)	1198 (34)	1196 (29)	1173	1183	1182	24	25	−23	−13	−14
21	1264	1251.5	4	(48%) ν(C2-N3)+22% ν(C-N9) +17% $\delta(NH_2)$+13% δ(ring)	1237 (11)	1244 (5)	1215	1225	1221	20	17	−29	−19	−23
22	1359	1355.5	8	(37%) δ(C6-H) + (34%) ν(C4-N9) + (22%) ν(ring)	1324 (17)	1320 (16)	1307	1316	1313	39	37	−13	−4	−7
23	1443	1441	12	(42%) δ(N1-H)+ (30%) ν(C-N3-C) +(22%) ν(N1-C6)	1382 (9)	1382 (17)	1387	1396	1398	61	63	5	14	16
24	1508	1499.9	23	(40%) ν(C-N9)+(25%) δ(C6-H) +(20%) δ(ring)+(15%) $\delta(NH_2)$	1441 (47)	1439 (51)	1450	1458	1457	69	67	11	19	18

(*continued*)

Table 5.3: *(continued)*

No.	Calculated				$\nu^{exp., c}$		Scaled[d]			Absolute error ($\nu^{cal.} - \nu^{exp.}$) in cm^{-1}				
										Calculated[e]		With a factor[f]	With LSE[g]	
	$\nu^{cal.}$	$\nu^{cal., a}$	*A*	Characterization[b]	Ne	Ar	(a)	(b)	(c)	(d)	(e)	(f)	(g)	(h)
25	1570	1562.2	21	(55%) ν(C4-C5) + (32%) ν(ring) + (13%) $\delta(NH_2)$	1540 (18)	1539 (21)	1509	1517	1516	31	29	−30	−22	−23
26	1637	1632.6	17	(90%) $\beta_s(NH_2)$	1569 (18)	1570 (14)	1574	1581	1584	67	68	4	11	14
27	1692	1681.4	64	(50%) ν(C=C) + (30%) ν (N3=C4) + (12%) ν(ring)	1625 (100)	1622 (91)	1627	1634	1631	70	65	5	12	9
28	1774	1758	100	(80%) ν(C=O) + (20%) ν(ring)	1725 (58)	1718 (100)	1706	1712	1708	56	49	−12	−6	−10
29	3211	3191.7	0	(75%) ν(C6-H) + (25%) ν(C5-H)			3087	3086	3090					
30	3236	3216.8	0	(75%) ν(C5-H) + (25%) ν(C6-H)			3111	3110	3116					
31	3610	3590.6	11	(98%) $\nu_s(NH_2)$	3474 (33)	3471 (12)	3471	3467	3473	139	129	0	−4	2
32	3629	3616.1	10	(98%) ν(N1-H)	3575 (36)	3564 (18)	3489	3486	3499					
33	3752	3719.3	6	(100%) $\nu_{as}(NH_2)$	3618 (50)	3600 (18)	3607	3603	3603	152	135	7	3	3
									Rms[h]	53.5	49	17.2	13.5	12.0

[a] With the 6-311+G(2d,p) basis set. [b] Abbreviations: ν, stretching; δ, in-plane bending; γ, out-of-plane bending. [c] Ref. [33]. [d] Wavenumbers scaled: (a) with the scale factor of 1.0013 for calculated wavenumbers lower than 800 cm^{-1}, and the scale factor of 0.9614 for higher wavenumbers [31]. (b) With the corresponding scale equations of Table 5.2. (c) With the 6-311+G(2d,p) basis set and with the corresponding scale equation of Table 5.2. [e] Absolute error in the calculated wavenumbers: (d) from the second column. (e) From the third column. (f) Error calculated from column (a). [g] Errors in the scaled wavenumbers: (g) From column (b). (h) From column (c). [h] Rms, defined as $(\sum (\nu^{cal.} - \nu^{exp.})^2/n)^{1/2}$, where the sum is over all the modes, *n*, and $\nu^{exp.}$ corresponds to the Ar matrix values.

C1, C2a, C2b, C3a and C3b of Cy. When the scaled wavenumbers of these tautomers are compared with the experimental values reported in an Ar matrix, it was not possible to identify the characteristic bands corresponding to the *enol* (C2a, C3b) or *imino* (C3a, C3b) tautomers. Thus, only the *keto* form C1 appears clearly in gas phase. However, in other studies, both C1 and C2 tautomers have been detected with a small amount of C3 [35].

In addition, with the help of computational methods, it is possible to modify the assignment reported by other authors. Thus, in Cy, the calculated C2=O stretch (scaled at 1,712 cm^{-1}, ninth column) is predicted as the strongest IR band in accordance with that observed in an Ar matrix spectrum at 1,718 cm^{-1} [34]. However, Radchenko et al. [36] pointed out that the most intense IR band in an Ar matrix at 1,620 cm^{-1} corresponds to the ν(C=C) stretch. We have corrected all errors in Table 5.3. For the discussion of the remaining bands, Table 5.3 is self-explanatory.

5.5.2.3 Assignment of all the bands in the uracil molecule

Table 5.4 shows the results obtained in the U molecule with the corresponding scaling equations of Table 5.2. The first column refers to the notation used for the ring normal modes [8, 27]. The main characterization of these modes appears in the second column. The third and fourth columns list the calculated wavenumbers with the B3LYP method and the 6-31G(d,p) and 6-311+G(2d,p) basis sets, respectively. The column with relative intensities was obtained by dividing the computed values by the intensity of the strongest line. Although wavenumber calculations on uracil have been determined by other authors using MP2 [37] and B3LYP DFT methods [38, 39], the values shown here represent the most accurate today.

Unfortunately, few studies have reported the gas-phase vibrational spectrum of U. The experimental values selected in the seventh to ninth columns were those reported in an argon matrix [40, 41] and in the gas phase [40, 42]. For the determination of the scaling equations, the values reported in the gas phase by Colarusso et al. [42] were selected as reference, since the assignments given there correspond most closely to our own. These wavenumbers appear underlined in Table 5.4. In the lack of these data, the values obtained in an Ar matrix were used. The scaling equations obtained in this way and listed in Table 5.2 have a good transferability to uracil derivatives [27]. With this correction (scaling) of the calculated wavenumbers, in general, they are remarkably close to the experimental wavenumbers, and thus they can be used for the assignments.

In the assignment of the uracil modes, Table 5.4 is self-explanatory, and only a few comments need be made: The N-H stretches appear as pure modes and with strong intensity, and thus they were clearly characterized and assigned. The N-H in-plane bendings are more complex because they appear mixed with other modes. These N-H modes have noticeable contributions to seven calculated vibrations. The description of mode 22 is complex. Mainly, it was characterized as δ(C6-H)+ δ(N-H)

Table 5.4: Comparison of the experimental wavenumbers (ν^{exp}, cm^{-1}) and calculated harmonic values ($\nu^{cal.}$, cm^{-1}) with the 6-31G(d,p) basis set, together with their relative (*A*, %) infrared intensities, and characterization obtained in the normal ring modes of uracil molecule with several methods. Absolute error (Δ) obtained in the calculated and scaled (ν^{scal}) wavenumbers.

No.	Characterization[a]	B3LYP			Infrared[c]			Absolute error $\Delta(\nu^{cal.} - \nu^{exp})$ in cm^{-1}			B3LYP[b,f]	
		ν^{cal}	$\nu^{cal.,\,b}$	*A*	Ar matrix[d]	Gas[d]	Gas[e]	B3LYP	B3LYP[b]	B3PW91	ν^{scal}	$\Delta(\nu^{scal} - \nu^{exp})$
1	Puckering N3	150	147	0							171.1	-
2	Puckering N1	170	165	0	185 w	185 w		−15	−20	−14	188.2	3
3	δ(OCNCO)	385	387	4	391m	377m	374 vw	−6	−4	−6	400	9
4	γ(C=C-H12)	396	395	4	411m	411m	395 w	1	0	1	407.6	13
5	δ(ring)	519	523	4	516.5m	527m	512 w	7	11	6	529.7	18
6	δ(ring) + δ(C=O)	541	543	1	536.4m			5	7	6	548.8	13
7	δ_{as}(ring) + δ(C=O)	558	547	1	559 w	588 w		−1	−12	0	552.6	−6
8	γ(N1-H)	563	560	8	551.2m	556m	545 w	18	15	25	565	20
9	γ(N3-H)	687	677	14	662.1 s	633m	659.5 w	27	17	32	676.6	18
10	γ(C4=O) + γ(C5-H)	729	728	2	718 w		717.4 vw	12	11	14	725.2	8
11	γ(C2=O)	752	765	10		769 s	756.5 w	−5	8	2	760.5	4
12	ν(ring)	772	769	0	759.2 sh?			−10	−13	−3	764.3	5
13	γ(C5-H) + γ(C4=O)	813	819	10	804 s	810 s	802 w	11	17	12	812	10
14	ν(C-C) + δ(N-H)	965	962	1	958.3 w	946 vw	952 w	13	10	18	948.4	−4
15	γ(C6-H)	970	972	0	963 w	974 w	972 sh	−2	0	1	957.9	−14
16	δ(NCC)	990	994	1	982 w	999 w?	990 sh	0	4	3	978.9	−11
17	ν(ring) + δ(C5-H)	1091	1086	1	1073 w	1089 s?	1082m	18	13	24	1066.7	−6
18	ν(C-N)+δ(C6H,N1H)	1198	1191	14	1184 vs		1172 s	26	19	37	1166.8	−5
19	δ(C5-H) + δ(N-H)	1231	1227	1	1217.4 w	1228m		44	40	50	1201.2	−16

(continued)

Table 5.4: (*continued*)

No.	Characterization[a]	B3LYP			Infrared[c]			Absolute error $\Delta(\nu^{cal.} - \nu^{exp})$ in cm^{-1}			B3LYP[b,f]	
		ν^{cal}	$\nu^{cal., b}$	*A*	Ar matrix[d]	Gas[d]	Gas[e]	B3LYP	B3LYP[b]	B3PW91	ν^{scal}	$\Delta(\nu^{scal} - \nu^{exp})$
20	δ(N3-H) + δ(C-H)	1382	1382	3	1359.3 vw	1360m	1356 sh	26	26	31	1349	−7
21	ν(C-N) + δ(N3-H)	1407	1406	20	1388.7 vs	1380 s	1387 s	20	19	28	1371.9	−15
22	δ(C6-H) + δ(N-H)	1422	1423	3	1399.6 vs	1396 ms	1400 s	22	23	26	1388.1	−12
23	δ(N1-H) + ν(N1-C)	1506	1498	18	1472ms	1480 s	1461 s	45	37	57	1459.6	−1
24	ν(C=C)	1690	1670	12	1644m	1632 vs	1641 s	49	29	60	1623.7	−17
25	ν(C4=O)	1808	1757	100	1741 vs	1688 vs		67	16	85	1706.7	−34
26	ν(C2=O)	1845	1791	92	1757.5 vs	1734 vs	1756 vs	89	35	108	1739.1	−17
27	ν(C6-H)	3221	3200	1		3076 w		145	124	155	3083	7
28	ν(C5-H)	3264	3242	0		3101 w	3124m	140	118	154	3123.1	−1
29	ν(N3-H)	3620	3592	11	3434.5 s	3427 w	3436 s	184	156	**206**	3456.9	21
30	ν(N1-H)	3658	3636	17	3484.3 s	3450 w	3484 s	174	152	197	3498.9	15
							Rms[g]	66	54	75		13.7

[a] Abbreviations: ν, stretching; δ, in-plane bending; γ, out-of-plane bending. [b] With the 6-311+G(2d,p) basis set. [c] Notation: sh, shoulder; vs, very strong; s, strong; m, medium; w, weak; vw, very weak. [d]Refs. [40,41]. [e]Ref. [42]. [f]With the scaling equations. [g]Rms, defined as $(\sum (\nu^{cal.} - \nu^{exp.})^2 / n)^{1/2}$, where the sum is over all the modes, *n*, and $\nu^{exp.}$ corresponds to the values underlined.

with some contribution of ν(C-N), in agreement with the assignment reported by Harsányi et al. [40] and Colarusso et al. [42]. Mode 21 is also complex and it has been characterized as ν(C-N)+ δ(N3-H) in agreement with Harsányi et al. [40] but in contrast to Colarusso et al. [42] and Lés et al. [43].

In the gas phase, the γ(N1-H) out-of-plane bending mode is clearly related to the band at 556 cm^{-1} [40] or at 545 cm^{-1} [42], in disagreement with Aamouche et al. [37], who related this band to the two calculated wavenumbers corresponding to δ(C2=O) and γ(N1-H) modes. The assignment of the very weak gas-phase band at 672 cm^{-1} [40] or at 692 cm^{-1} [42] should be changed and related to the γ(N3-H) mode.

The calculated C2=O and C4=O stretches are the strongest IR bands, and they are predicted with similar intensity. The scaled wavenumbers of both modes are at 1,739.1 and 1,706.7 cm^{-1} (fourteenth column in Table 5.4), modes C2=O and C4=O, respectively, in good agreement with the gas-phase results. The C=O out-of-plane bending appears in general mixed with γ(C-H) modes. The description of mode 13 as γ(C5-H)+ γ(C4=O) is in agreement with Colarusso et al. [42] and other authors [38, 43, 44]. The band reported in the gas phase [42] at 717.4 cm^{-1} and in an argon matrix at 707.4 and 719 cm^{-1} by Ivanov et al. [41] and Szczepaniak et al. [38], respectively, were related here to mode 10, defined as γ(C4=O) + γ(C5-H).

As we show here, with accurate scaling, it is possible to carry out a good assignment and in some cases to correct that reported by other authors.

5.5.2.4 Scaling in Cy and U molecules: accuracy of the different methods

The accuracy of the two scaling procedures mentioned above on the Cy molecule and the two theoretical methods used is shown in Table 5.3. The absolute error obtained in the wavenumbers is shown in the last five columns. They were calculated with the experimental values in an Ar matrix. In the last row, the root mean square (rms) error of each level and procedure is determined. With the LSE procedure, the errors obtained in the predicted wavenumbers were very small; the mean deviation was 13 cm^{-1} (1 %). The values of these rms errors were very close to those obtained in other related compounds studied by us [9, **45**, **46**]. This good match of scaled to experimental wavenumbers helps in the assignment and analysis of the experimental fundamental modes.

Similar to the Cy molecule, the absolute errors obtained in the calculated and scaled wavenumbers of the U molecule are shown in Table 5.4. The largest values correspond to the calculated wavenumbers and they are marked in bold type. However, with the scaling equation they are remarkably reduced as shown in the last column of the Table. The bottom of the table shows the rms error obtained for the calculated and scaled wavenumbers at several computational levels.

Calculations at other levels were also carried out on the U and Cy molecules. The errors obtained in the U molecule and by the two scaling procedures are shown in Table 5.5. As can be seen, remarkable differences appeared between the HF and B3LYP methods. In both cases, the large rms error in the calculated wavenumbers is remarkably reduced with the scaling, especially by HF. However, the error obtained

Table 5.5: Errors obtained in the calculated and scaled wavenumbers of the uracil modes by the different procedures and methods.

Method	Calculated wavenumbers			Scaled wavenumbers with an overall factor			Scaled wavenumbers with LSE		
	rms	Largest error (cm^{-1})		rms	Largest error (cm^{-1})		rms	Largest error (cm^{-1})	
		Positive	Negative		Positive	Negative		Positive	Negative
HF/6-31G(d,p)	184	427	6	23	53	37	22.6	46	57
HF/6-31++G(d,p)	177	418	11				16.7	27	50
MP2/6-31G(d)	82	187	44	37	53	95	25.4	50	66
BP86/6-31G(d,p)	35	86	44	33	56	50	18.1	34	32
BLYP/6-31G(d,p)	34	73	49	34	54	54	19.6	34	36
B3P86/6-31G(d,p)	77	207	14	24	46	44	15	32	26
B3LYP/6-31G(d,p)	66	184	15	21	44	40	13.8	24	23
B3LYP/6-311+G(2d,p)	54	156	20				13.7	21	34
B3PW91/6-31G(d,p)	75	206	14	25	50	41	14.9	30	26
M06-2X/6-31G(d,p)	79	192	13				19.1	53	31
M06-L/6-31G(d,p)	70	190	13				19	49	25

after scaling by HF continues to be higher than by B3LYP. That is the HF method should not be used for assignment of the experimental wavenumbers.

With the use of a scaling equation, a remarkable reduction of the error is obtained by B3LYP and thus of the risk to a mistake. The best for this purpose is the 6-311+G(2d,p) basis set shown in the last three columns of Table 5.5. Larger basis sets represent an excessive increase in the computational cost for a very slight improvement. The largest absolute error corresponds to vibration no. 12, a ring bending. Large errors also appear in the vibrations n° 21 and 25, corresponding to C-N and C-C stretches, respectively.

Finally, it can be concluded that the LSE procedure leads to very low errors, and lower than using an overall scaling factor. With B3LYP, the wavenumbers are significantly close to the experimental and much better than by HF. This B3LYP method is also the best among the DFT methods used, and it is the recommended method for carrying out an assignment of the vibrational bands.

5.5.3 Identification of the tautomers present in the isolated state

5.5.3.1 Tautomerism in nucleic acid bases

The nucleic acid bases can undergo *keto-enol* tautomerism. Much of the interest of the tautomerism is due to the fact that tautomers induce alterations in the normal base pairing, leading to the possibility of spontaneous mutations in the DNA or RNA helices. On the basis of geometry, ionization potential and dipole moment, it is virtually impossible to decide which arrangement is actually the most stable one. However, computational methods applied to the vibrational spectra may help in deciding the most stable arrangement (tautomer). The absorption bands due to C=O, NH and OH groups give the most straightforward information about the tautomeric forms present, because they correspond to the characteristic, well-localized vibrations of the functional groups directly involved in tautomeric changes. Thus, special attention was paid on these modes.

All the tautomers in the Cy molecule appear with relative energies much lower than their counterparts in the U molecule. Two features can explain it: (i) the negative charge on N1 atom is lower in Cy than in U, with longer N1-H bond length. (ii) The negative charge on the O2 atom is higher in Cy than in U, with longer C=O bond length. Both features favor tautomerism in Cy molecule. Thus, there is a special interest to study this molecule and its derivatives [15, 16, **47**].

The Cy molecule has been studied intensively by IR spectroscopy in the free monomeric form (in low-temperature inert matrices [34, 36]) and in the crystalline phase at room [48] or at low temperatures [49]. The different aggregate states can contain two tautomeric forms [33, 50]: the gas phase with both the *enol* and the *keto* forms, whereas in the crystalline state and polar solvents only the *keto* form has been observed [34]. For its interpretation, the theoretical vibrational spectrum of Cy has been predicted at different levels of approximation [34, 51].

5.5.3.2 Tautomerism in 5-bromocytosine

Similar to the Cy molecule, 5-BrCy can exist in various tautomeric forms. However, this molecule has been much less studied than Cy, and thus it is discussed here. The computational methods have been applied to vibrational spectroscopy to simulate (scaled) the IR spectrum of the different tautomers of 5-BrCy. In the isolated state, the *enol* form C2b of 5-BrCy is the most stable one. The next most stable tautomer is the *enol* form C2a, 3.18 kJ/mol above C2b in 5-BrCy, and 12.96 kJ/mol above C2b in the Cy molecule with the MP2 method. The bromine atom in position 5 favors the tautomerism with much lower relative energies (about 66 %) in 5-BrCy than in Cy.

In the isolated state, the harmonic vibrational bands computed in the 5-BrCy ring are shown in Table 5.6. The second column lists the calculated wavenumbers with the 6-31G(d,p) basis set in tautomer C1. The third column collects their relative IR intensities (A) in percent. They were obtained by normalizing the computed value to the intensity of the strongest band. The assignment with the calculated percent potential energy distribution (PED) of the different modes for each vibration appears in the fourth column. Contributions lower than 10 % were not considered. The scaled wavenumbers in the fifth column correspond to tautomer C1 and those in the ninth column to tautomer C2b. With these data obtained theoretically, the experimental spectrum in an Ar matrix reported by Jaworski et al. [52] can be now analyzed in detail. For this task, the scaled values were directly compared to the experimental values, and from this comparison it was possible to separate those corresponding to tautomer C1 (sixth column), from those to tautomer C2b (tenth column). In Table 5.6, only the experimental bands with high intensity are shown, and those selected in the comparison with the scaled values are underlined.

Through the comparison of these spectra, we have identified, separated and assigned the bands corresponding to the different tautomers, which in this case are only the amino-oxo form (tautomer C1) and the amino-hydroxy form C2b which is the most stable form in accordance to the computations, Figure 5.5. Due to many bands between both conformers appear very close in frequency, the IR intensity criterion was followed in these cases. Thus, the assignment reported previously by Jaworski et al. [52] has been improved by using scaling, and all bands accurately assigned to both tautomers. Bands corresponding to tautomer C3b, simulated theoretically, were not found in the experimental spectrum.

5.5.4 Simulation of the crystal unit cell of a compound and the interpretation of its vibrational spectra

Figure 5.6 shows the optimized simulated tetramer form of the Cy molecule obtained from the crystal unit cell reported by X-ray [53]. The molecules adopt their amino-oxo tautomeric form, in which they are associated to form ribbons stabilized by N-H···O hydrogen bonds that involve N1-H and NH_2 groups and the carbonyl oxygen atom. Four molecules appear in the crystal unit cell.

Table 5.6: Comparison of the calculated harmonic wavenumbers (ν^{cal}, cm^{-1}), relative infrared intensities (A, %), scaled harmonic wavenumbers (ν^{scal}, cm^{-1}), experimental IR values ($\nu^{exp.}$), and characterization obtained in gas phase of both tautomers of 5-BrCy molecule at the B3LYP/6-31G(d,p) level.

No	Tautomer C1					Tautomer C2b			
	ν^{cal}	A	Characterization	$\nu^{scal,\ a}$	Exp.[b]	ν^{cal}	A	$\nu^{scal,\ a}$	Exp.[b]
1	3630	9	(95%) ν(N1-H)	3464	3454, 3441				
2	3228	0	(99%) ν(C6-H)	3084		3189	1	3047	
3	1822	100	(79%) ν(C=O) + (14%) δ(N1-H)	1756	1752, 1724, 1718				
4	1696	49	(41%) ν(C=C) + (38%) ν(C-N) + (12%) $\beta_s(NH_2)$	1637	1629, 1618, 1609	1635	5	1579	1588
5	1556	19	(49%) ν(N3-C4) + (34%) ν(N1-C-C) + (11%) $\beta_s(NH_2)$	1505	1507	1598	39	1544	1556
6	1499	17	(48%) ν(N3CC5) + (24%) $\delta(NH_2)$ + (20%) ν(N1-C6)	1451	1434	1489	70	1441	1442, 1437
7	1435	7	(40%) δ(N1-H) + (33%) ν(N-C4)+(16%)Δ(ring)	1390	1420, 1418				
8	1331	0	(49%) δ(C-H) + (29%) ν(CCC) + (12%) δ(N1-H)	1292	1286	1335	6	1295	1300, 1296
9	1269	3	(42%) ν(C2-N3)+ (25%)ν(CN10)+(21%) Δ(ring)	1233	1242, 1234, 1220	1312	3	1274	1258
10	1197	7	(72%) δ(H-N1-C6-H) + (17%) ν(ring)	1165	1178, 1171				
11	1039	6	(94%) ν(ring)	1016		1057	6	1033	1055
12	937	1	(91%) γ(C6-H)	920		977	1	957	950
13	916	1	(65%) ν(ring) + (35%) $\delta_{as}(NH_2)$	900		988	3	967	976
14	772	4	(84%) γ(N-CO-N) mainly in C2 + (16%) Γ(ring)	764	773	799	7	789	789
15	764	1	(74%) ν(ring) + (26%) $\delta_s(NH_2)$	756		795	1	785	779
16	745	0	(66%) γ(C4-C)+(21%) Γ(ring) + (13%) $\tau(NH_2)$	738		731	0	725	
17	627	2	(67%) ν(ring) + (21%) $\delta_{as}(NH_2)$ + (12%) ν(C-Br)	627		650	2	648	652
18	625	8	(71%) γ(N1-H) + (16%) Γ(ring) +(13%) $\tau(NH_2)$	625	611, 605				
19	576	1	(45%) Δ(ring) + (33%) δ(C=O) + (22%) $\delta(NH_2)$	579		581	1	583	601

(continued)

Table 5.6: (*continued*)

No	Tautomer C1					Tautomer C2b			
	ν^{cal}	*A*	Characterization	$\nu^{scal,\,a}$	Exp.[b]	ν^{cal}	*A*	$\nu^{scal,\,a}$	Exp.[b]
20	536	0	(70%) Δ(ring) mainly in N1, N3 + (30%) δ_{as} (NH_2)	541	545	531	1	536	<u>530</u>, 526, 518
21	396	1	(55%) γ(N1-CH=C) + (30%) Γ(ring) + (15%) τ (NH_2)	409		455	0	464	
22	389	0	(36%) δ(C=O) + (28%) Δ(ring) + (26%) δ_{as} (NH_2)	402		369	2	383	
23	282	0	(55%) δ(C-Br) + (29%) Δ(ring) + (12%) δ(NH_2)	301		282	1	301	
24	255	0	(85%) puckering N1 +(15%) τ(NH_2)	275		300	1	318	
25	205	10	(51%) puckering N3 + (49%) γ(NH_2)	228		220	3	242	
26	204	1	(41%) δ(C-Br) + (31%) Δ(ring) + (28%) δ(NH_2)	227		201	0	224	
27	77	1	(30%) Γ(ring) + (46%) γ(C=O) + (24%) γ(NH_2)	107		84	18	113	

[a]With the scaling equation: $\nu^{scaled} = 34.6 + 0.9447 \cdot \nu^{calculated}$, ref [8]. [b] IR in Ar matrix, ref [52]. The highest IR intensity is shown underlined.

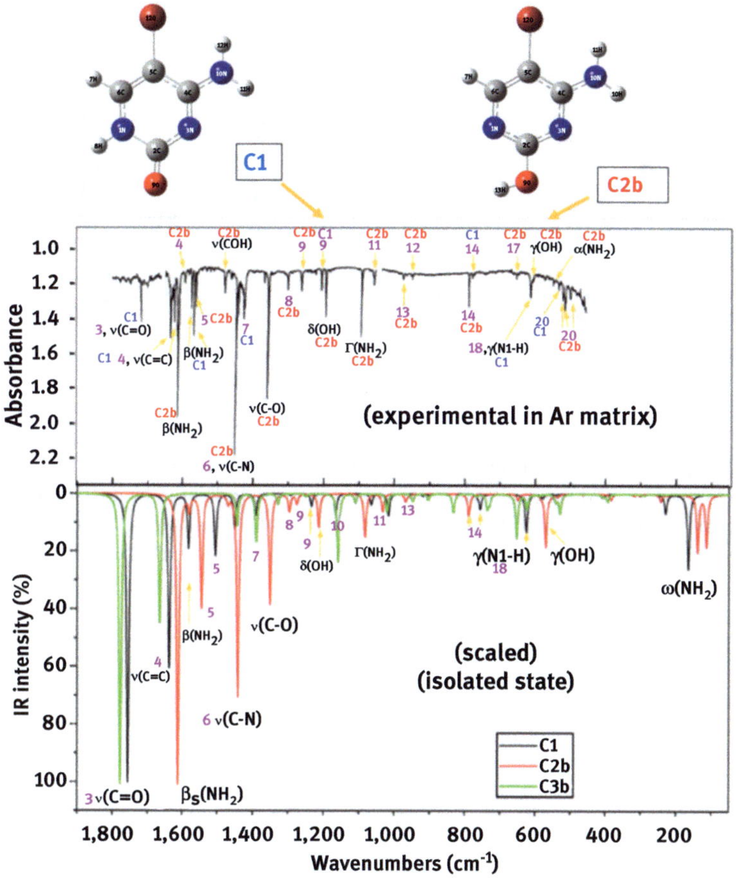

Figure 5.5: Assignment to tautomers C1 and C2b of the experimental IR spectrum in argon matrix of 5-bromocytosine.

Unfortunately, the detailed analysis of this structure has not yet been reported. Thus, the molecule of 5-chlorocytosine (5-ClCy) has been selected as example.

5.5.4.1 5-Chlorocytosine

The crystal unit cell of 5-ClCy was simulated at the B3LYP/6-31G(d,p) level by a tetramer form, Figure 5.7. Its simulated spectrum appears closer to the experimental spectrum in the solid state than the simulated spectrum of the monomer form. Figure 5.8 only shows the comparison between the experimental (IR) and that scaled with the tetramer form. Similar agreement is observed with the Raman spectrum.

The scaled and experimental wavenumbers in the monomer and tetramer forms are collected in Table 5.7, together with the corresponding assignment.

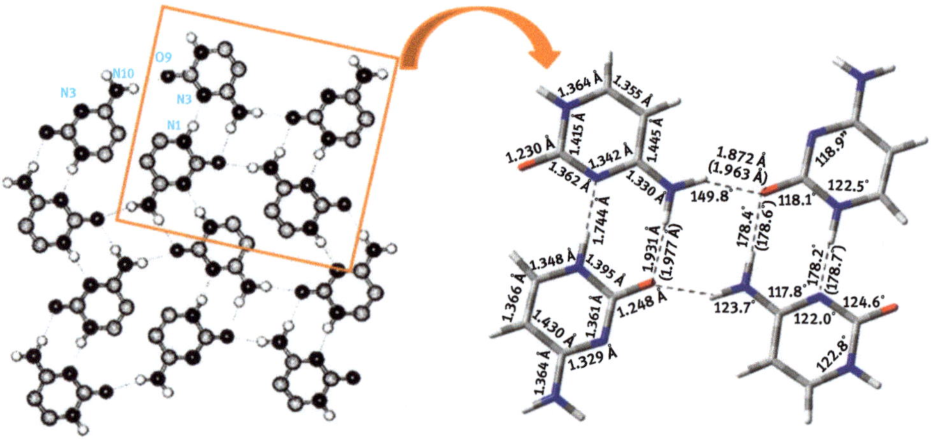

Figure 5.6: Schematic representation of the crystal structure of cytosine in the solid state [16, 53]. Hydrogen bonds are formed by all three available protons and they are indicated by dashed lines. At the right side appears the simulation at the B3LYP/6-31G(d,p) level of the crystal unit cell of Cy through a tetramer form. In parentheses appears the values obtained at the O3LYP/6-31G(d,p) level.

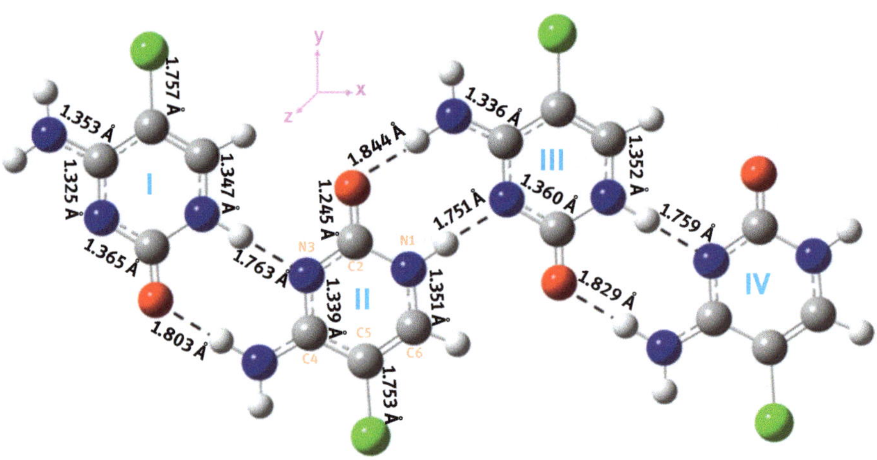

Figure 5.7: Simulation of the crystal unit cell of 5-chlorocytosine by a tetramer form at the B3LYP/6-31G(d,p) level.

They are the most accurate today for this molecule. The scaling equation $\nu^{\text{scaled}} = 4.8 + 0.9671 \cdot \nu^{\text{calculated}}$ deduced from the Cy molecule (Table 5.3) at the B3LYP/6-311+G(2d,p) level was used for the monomer, while $\nu^{\text{scal}} = 34.6 + 0.9447 \cdot \nu^{\text{cal}}$ at the B3LYP/6-31G(d,p) level was used for the tetramer. Four wavenumbers appear in the tetramer form for each characterized vibration corresponding to its four

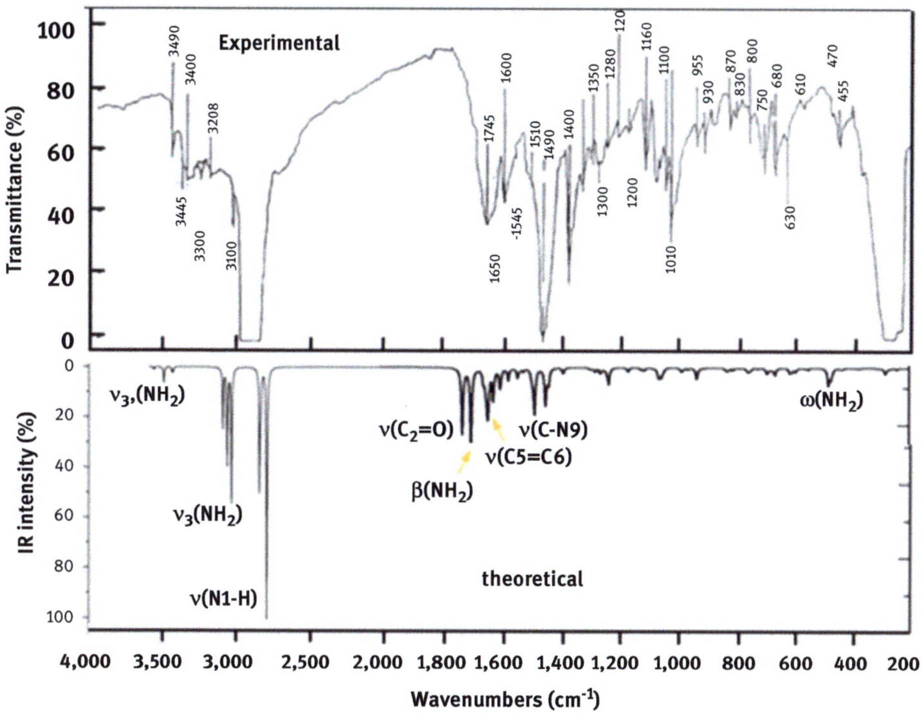

Figure 5.8: Experimental and theoretical IR spectra of 5-chlorocytosine in the 4,000–200 cm^{-1} range. The predicted scaled spectrum was obtained using the LSE procedure in the tetramer form at the B3LYP/6-31G(d,p) level.

molecules. Of these four wavenumbers, those determined in the molecules II and III (see Figure 5.7) are underlined. These wavenumbers appear affected by the internal H-bonds of the crystal, and they represent better the solid state. Among these wavenumbers, that with the highest IR intensity appears in bold type, while that with the highest Raman intensity is shown in italic type. The main characterization carried out, and corresponding to the tetramer form, is shown in the last column.

The absence of a ν(OH) band in the 3,500–3,700 cm^{-1} range and the appearance of ν(C=O) modes as strong bands indicate that in the solid state this compound is only in the *keto* form. Another feature is that the majority of the fundamental modes absorb in relatively narrow wavenumber ranges. We have changed the assignments of several bands reported by other authors [20] according to our scaled values. A good match between the scaled wavenumbers in the tetramer to the experimental ones was found, and much better than that with the monomer form. That is the simulation of the crystal unit cell is appropriate for an accurate assignment of the experimental bands in the solid state.

Table 5.7: Comparison of the experimental wavenumbers (cm^{-1}) and harmonic calculated wavenumbers (ν^{cal}, cm^{-1}), relative infrared intensities (A, %), relative Raman scattering activities (S, %), scaled wavenumbers (ν^{scal}, cm^{-1}), and characterization obtained in 5-ClCy molecule calculated at B3LYP/6-311+G(2d,p) level.

Monomer				Tetramer	Exp.[b]		Characterization
ν^{cal}	A	S	ν^{scal}	$\nu^{scal,\ a}$	IR	Raman	
3731.4	9	23	3602.2	3590, 3523, ***3521***, 3520	3451 s	3448 ms	ν_{as}(N-H) in NH_2
3611.7	9	100	3487.3	3468, 2872, 2864, ***2822***	3320 w	3320 ms	ν(N1-H)
3599.9	13	74	3476	3457, 3121, 3093, ***3063***	3182 w	3174m	ν_s(N-H) in NH_2
3207.9	0	54	3099.8	***3091***, 3085, 3084, 3081	3090 w	3081m	ν(C6-H)
1761.1	100	25	1711.3	1738, **1707**, 1686, *1666*	1708 w	1708 ms	ν(C2=O)
1678.5	46	6	1632.1	**1633**, *1610*, 1608, 1608	1630 b	1626 ms	ν(C5=C6) + ν(ring) + $\beta_s(NH_2)$
1627	20	7	1582.6	1658, *1653*, **1650**, 1582	1655 ms	1651 ms	$\beta_s(NH_2)$ + ν(C-N)
					1610 s	1604 s	
1542.6	19	12	1501.6	**1550**, 1540, 1530, *1500*	1525 ms	1538 ms	ν(CCCN3) + δ(N1-H)
1485.1	24	4	1446.4	1503, *1496*, **1492**, 1455	1510 ms	1512m	ν(C-N10) + ν(NC6) + $\beta_s(NH_2)$ +δ(CH)
1430.7	8	5	1394.2	1454, 1450, ***1445***, 1395	1472 s	1470 s	ν(N1C6) + δ(N1-H) +ν(C4C5)+ $\beta_s(NH_2)$
					1405 ms	1394 w	
1325.9	0	7	1293.7	1303, ***1302***, 1300, 1297	1342 s	1342m	δ(C-H) + δ(N1-H) + ν(C-N10)
1256	3	9	1226.6	1285, 1284, ***1267***, 1256	1295 s	1294m	ν(C2-N3) + ν(C-N10) + ν(ring)
1195.6	8	2	1168.6	*1245*, 1240, **1237**, 1170	1227 s	1220 s	δ(N1-H) + δ(C-H)
1092.7	3	1	1069.9	1127, 1126, **1116**, *1086*	1128 vs	1130m	r(NH_2) + ν(ring)
1060.9	8	3	1039.3	*1067*, 1063, **1056**, 1050	1055 vs	1060 w	ν(ring) + ν(C5=C6)
					1034 vs	1028 vs	
932.6	1	0	916.2	**937**, *935*, 934, 908	890 ms	892 ms	γ(C6-H)
916.9	1	1	901.1	***965***, 958, 939, 922	972 w	962 ms	δ(ring) + r(NH_2)
780.4	4	0	770.1	763, ***755***, 754, 748	740 s	748m	γ(C=O) + γ(ring)

(*continued*)

Table 5.7: (*continued*)

Monomer				Tetramer	Exp.[b]		Characterization
ν^{cal}	*A*	*S*	ν^{scal}	$\nu^{scal,\ a}$	IR	Raman	
771.9	0	15	762	***788***, 785, 780, 774	785 w	785m	ν(ring) + ν(N1-C) + $\delta(NH_2)$
750.7	0	0	741.6	**725**, *724*, 723, 722			γ(C4) + γ(C=O) + γ(ring)
658.5	3	2	653.2	*692*, 685, 676, **666**	652 ms	654 vs	ν(C-*Cl*) + ν(ring)
612.3	9	0	608.8	989, 986, 981, ***615***	982 w		γ(N1-H)
583.4	1	5	581.1	608, *604*, 598, **596**	620 w	618m	δ(C=O) + δ(ring) + r(NH_2)
					595 ms	588 ms	
562	0	0	560.6	**834**, 825, 810, *565*	820 w		$\tau(NH_2)$
542	0	2	541.4	592, 576, 559, ***551***	525 ms	540 ms	δ(ring) + δ(C=O) + r(NH_2)
410.2	0	4	414.9	**467**, 453, 437, *431*	435 s	438m	δ(C-*Cl*, ring) + r(NH_2)
397.5	1	0	402.7	442, **441**, 438, *411*			γ(CCN) + γ(N1-H) + $\tau(NH_2)$
352.4	0	2	359.4	*391*, 387, 383, **375**	395m	390 vw	δ_{as}(NC4C*Cl*)+δ(C=O)+δ(ring)+r(NH_2)
267.1	0	0	277.5	310, **308**, 299, *295*			γ(C-*Cl*) + γ(ring) + $\tau(NH_2)$
240.9	1	1	252.4	**285**, 283, *277*, 267	290 b	295 ms	δ_s(N-C4-C-*Cl*) + r(ring)
205.1	4	0	218	**243**, 238, *233*, 219	251 ms	250 w	γ(N3) + $\omega(NH_2)$ + γ(C=O)
123.5	20	0	139.7	***480***, 479, *473*, 156	472m	482 ms	$\omega(NH_2)$
80.8	0	0	98.7	***153***, 144, 130, 128			γ(ring) puckering on O, *Cl*

[a]With the equation at the B3LYP/6-31G(d,p) level: $\nu^{scal} = 34.6 + 0.9447\ \nu^{cal}$. [b]Ref. [20].

5.5.4.2 5-Bromocytosine

X-ray data of 5-BrCy has not yet been reported, and thus the tautomer present in the solid state is not known, but it is possible to resolve this problem by simulation of the crystal unit cell of all the possible tautomers using as reference the structure reported in related molecules, which in the present case is 5-ClCy. The crystal data of this unit cell (the tetramer form) was the starting point used for the simulation of the solid state (**Figure 5.9**).

In the isolated,state the molecule of 5-BrCy is calculated to be full planar, and the small deviations from planarity observed in crystals of Cy [53] are due to the intermolecular H-bonds. Our simulations in the tetramer form also show these deviations,

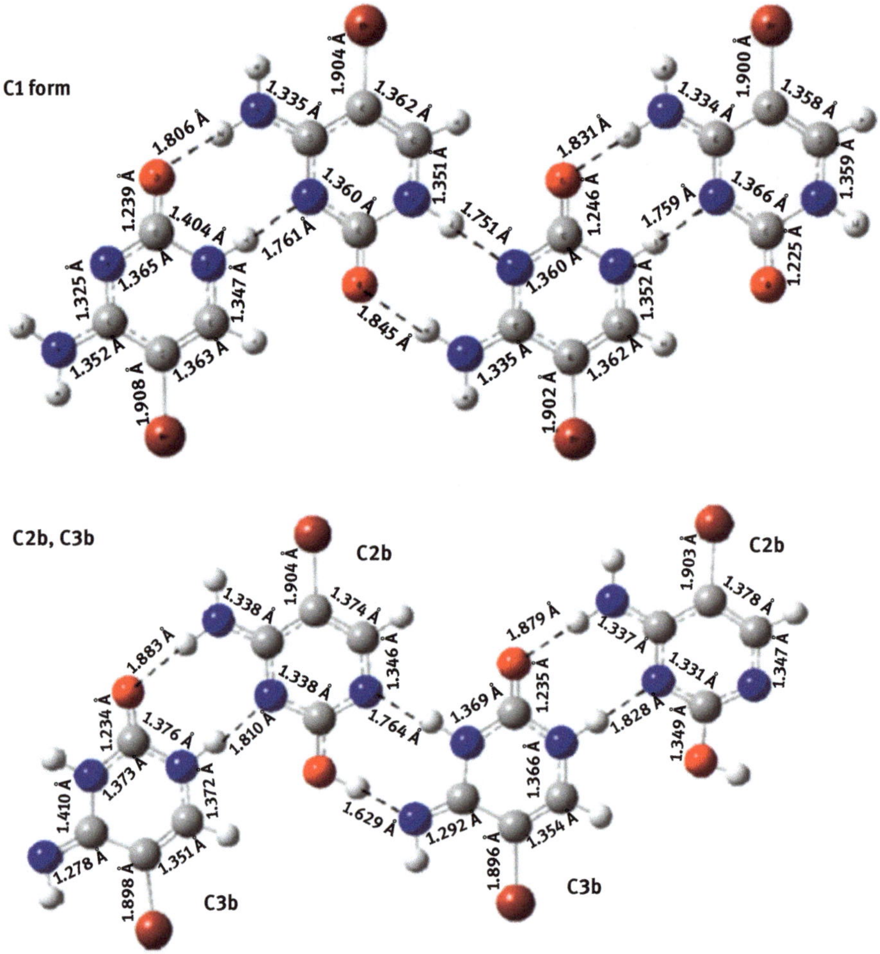

Figure 5.9: Simulation by a tetramer form of two possible crystal unit cells of 5-bromocytosine with three tautomeric forms.

in general with values slightly higher than in the crystal. Due to the orientation of the bromine atom in the tetramer form, and its almost zero interaction with other atoms, its bond length is almost the same as that in the isolated state. The simulations were carried out at the B3LYP/6-31G(d,p) level [54]. The tetramer form with tautomer C1 is 30.5 kJ/mol more stable than with tautomers C2b-C3b. The simulated spectra of these tetramer forms appear closer to the experimental spectrum in the solid state than the simulated spectrum of the monomer form. As an example of this concordance, the experimental (IR) and the scaled spectrum of this molecule are compared in Figure 5.10, for simplicity only in the dimer form and in the 3,700–2,700 cm^{-1} range. Similar agreement is observed in the other regions, as well as in the Raman spectrum.

The scaled values of the ring vibrations in two of the tetramer forms are collected in Table 5.8. Four wavenumbers appear for each vibration corresponding to the four 5-BrCy molecules of the tetramer. Of these four wavenumbers, that with the highest IR intensity is shown in bold type, while that with the highest Raman intensity is shown in italic type. These scaled tetramer values can be directly compared to the

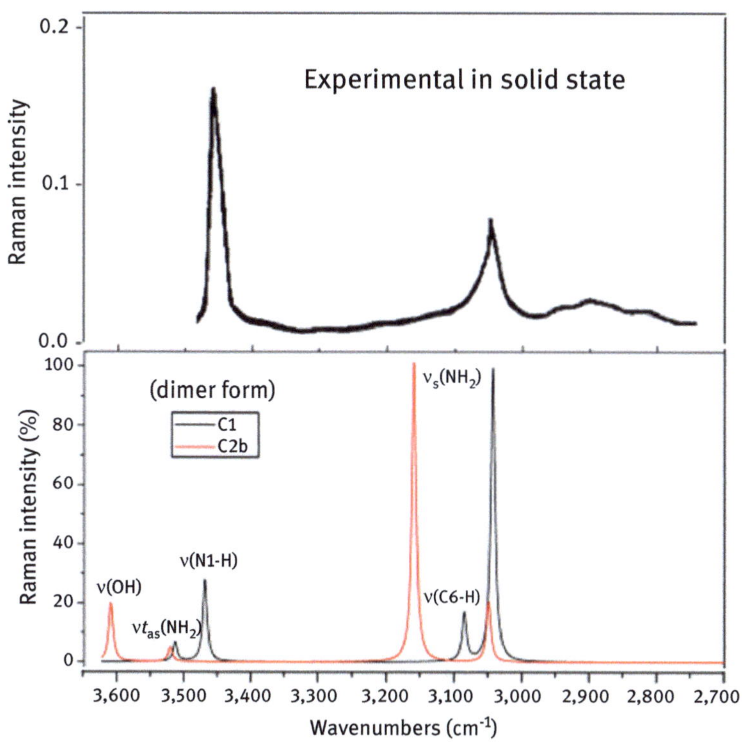

Figure 5.10: Experimental and theoretical scaled Raman spectra in the dimer form of 5-bromocytosine in the 3650–2700 cm^{-1} range and in the C1 and C2b tautomers using the scale equation $\nu^{scaled} = 34.6 + 0.9447\ \nu^{calculated}$.

Table 5.8: Comparison of the scaled harmonic wavenumbers (ν^{scal}, cm^{-1}) obtained in the tetramer forms of the ring of 5-BrCy with the experimental IR and Raman values, and the absolute error $\Delta(\nu^{scal} - \nu^{exp})$ (in cm^{-1}) obtained. The highest calculated IR intensity is shown in **bold** type while the highest Raman intensity is printed in *italic* type.

Tetramer (C1)		Tetramer (C2b and C3b)		Experimental in the solid state	
ν^{scal}	Δ	ν^{scal}	Δ	IR[a]	IR[b]
3466, 2870, 2859, ***2818***	−82	3459, *2969*, 2948, **2816**	−84	3210 w	2900 b,vs
3089, **3085**, 3084, 3082	15	3095, 3093, ***3054***, 3052	−16	3095 w	3070 vs
1737, **1705**, 1684, *1665*	−5	***1730***, 1726	20	1726 w	1710 vs
1633, *1606*, 1605, 1604	23	1662, ***1656***, 1617, 1612	46	1625 w, 1610 m	
1488, 1448, *1443*, 1394	16	***1593***, 1546, 1493, 1484	121	1510 w	1510 m
1497, *1496*, **1456**, 1452	−54	1539, *1465*, 1445, **1437**	−73	1472 vs, 1455 w	1433 s
1548, 1539, *1529*, 1493	−22	*1532*, 1528, **1522**, 1361	−48	1570 ms	1556 m
1320, **1303**, 1302, *1301*	−39	*1334*, **1330**, 1303, 1278	−12	1342 vs	1337 m
1284, 1284, **1266**, *1248*	−19	1421, **1352**, *1311*, 1280	67	1285 ms	1285 s
1270, *1243*, **1239**, 1169	9	1301, *1288*, **1231**, 1216	1	1230 m	1235 s
1044, 1041, **1034**, 1030	19	1050, ***1046***, 1042, 1021	31	1065 ms, 1015 s	1066 s, 1004 m
992, 946, ***943***, 920	−12	976, ***959***, 946, 943	4	975 ms, 955 ms	943 w
966, 957, 939, 928	53	***1001***, 998, 979, 965	88	900 w	913 m
766, ***760***, 758, 750	15	**783**, 780, 738, *736*	38		745 m
787, 784, 780, 773	9	792, *786*, 783, **775**	−3	776 ms	778 m
740, 737, *737*, **735**	10	***730***, 729, 724, 723	5	725 s	725 w
665, 659, 649, 641	19	673, 660, *647*, **633**	−13	632 m	646 vs
995, 988, *942*, **615**	15	964, **897**, 893, *655*	297		600 s
604, 598, *596*, 586	24	616, 608, 601, ***598***	18	580 m	
590, 573, 557, *551*	45	589, 567, 555, ***549***	4	545 ms	
451, 448, 444, ***416***	−3	*471*, **466**, 432, 419	47	430 ms	440 s, 419 w
458, 445, *430*, 415	3	**457**, 441, 422, *410*	2	495 w	470 m, 455 w

(*continued*)

Table 5.8: *(continued)*

Tetramer (C1)		Tetramer (C2b and C3b)		Experimental in the solid state	
ν^{scal}	Δ	ν^{scal}	Δ	IR[a]	IR[b]
313, 309, **308**, 304	28	***316***, 309, 304, 303	36	380 w, 280 b	-
303, **301**, 292, 289		**328**, *324*, 308, 305		-	-
244, *238*, 232, 219	−2	*247*, **240**, 201, 182	0	-	-
258, 255, *246*, 238	41	**278**, 257, *245*, 235	73	-	-
150, 142, **130**, *123*		***147***, 138, 130, 127		-	-

[a] Ref [20]. [b] Ref [54].

experimental IR and Raman data reported in the solid state [20, [54]. We have underlined the experimental bands selected for this comparison. The absolute error (Δ) between the observed and the scaled wavenumbers is lower in tetramer C1 than in C2b-C3b. I.e. The spectrum of crystalline solid 5-BrCy shows that this molecule exists in the crystal in only the amino-oxo form (C1).

The preference for only one tautomeric form in the crystal than in the isolated state is a clear indication of the importance of intermolecular interactions, in particular H-bonding, to determine the structure of the condensed phase. Because molecules in the crystal are involved in hydrogen bonding and in other intermolecular interactions, it is expected that these H-bonds significantly change the spectrum obtained in the solid state compared to that in the gas phase (or in Ar/Ne matrices). The bands corresponding to the stretching and bending vibrations of the N1-H, C=O and NH_2 groups change more drastically than those related to the ring or to the CH group. Thus, the calculations of the tetramer form help to interpret the spectra in the crystalline solid.

5.6 Summary and conclusions

i. Several examples of the applicability of the computational methods to analysing the vibrational spectra of cytosine, uracil, 5-chlorocytosine and 5-bromocytosine nucleic acid bases are shown. DFT quantum chemical methods were selected as the most appropriated, in particular B3LYP, which was the main method used here.
ii. The B3LYP/DFT method was used to characterize the ring normal vibration modes of uracil and cytosine molecules through the atomic displacement vectors involved in each calculated vibrational frequency. This application of the computational methods permits a clear identification and assignment of all the bands observed experimentally.

iii. An accurate simulation of the vibrational spectra of the molecules under study was performed using DFT methods. These theoretical/scaled spectra can be compared to the experimental IR and Raman spectra, and therefore all the bands can be assigned.

iv. To correct the systematic error of the theoretical methods, accurate scaling procedures are presented and used in the calculated spectra of the molecules under study. The use of the LSE scaling equation procedure significantly reduced the error in the calculated wavenumbers and it is the best for this purpose [55]. Thus, we recommend its use instead of the single overall factor procedure utilized in the literature.

v. With the help of DFT calculations, vibrational spectroscopy was used as a tool for the identification of the tautomers present in the isolated state. As an example, the study carried out on the 5-bromocytosine molecule is described. In the isolated state, at least the C1 and C2b tautomers are presented. The wavenumbers corresponding to these tautomers were identified and assigned in the IR experimental spectrum reported in an Ar matrix.

vi. Another example of this utility in the identification of tautomers is shown in the cytosine molecule. The scaled wavenumbers of the different tautomers were compared to the experimental values reported in Ar/Ne matrices. However, in this case it was not possible to identify the characteristic bands corresponding to *enol* or *imino* tautomers, Thus, only the *keto* form appears clearly in the spectrum of this phase, in contrast to that reported using other techniques.

vii. Computational methods were also used to simulate the crystal unit cell of 5-chlorocytosine and 5-bromocytosine nucleobases, with the consequent interpretation of their vibrational spectra. With this simulation, a remarkable improvement was reached in the accuracy for the assignment of their spectra in the solid state; in particular in the vibrations involved in intermolecular H-bonds. In 5-bromocytosine, two tetramer forms were calculated. The scaled wavenumbers of these tetramer forms appeared in accordance with the experimental IR and Raman data in the solid state, and thus all the normal modes were identified and discussed. Vibrational spectroscopy indicated that in the solid state the biomolecule 5-bromocytosine exists only in the amino-oxo form.

viii. The possibility of tautomerization of cytosine is calculated to be much more likely than that of 5-bromocytosine. The effect of the Br atom is to prevent tautomerization from taking place, and it is higher in 5-bromocytosine than in 5-bromouracil. All the tautomers in the cytosine molecule appear with much lower relative energies than their counterparts in the uracil molecule.

References

[1] Alcolea Palafox M. The prediction of vibrational spectra: The use of scale factors. Recent Res. Devel. in Physical Chem. Transworld Research Network: Trivandruw, India. Vol. 2 1998:213–232.

[2] Thirunarayanan S, Arjunan V, Marchewka MK, Mohan S, Atalay Y. Characterisation of 1,3-diammonium propylselenate monohydrate by XRD, FT-IR, FT-Raman, DSC and DFT studies. J Molec Struct. 2016;1107:220–230.

[3] Farasat M, Reza Shojaei SH, Maqsood Golzan MM, Farhadi K. Theoretical study of the potential energy surface and electric dipole moment of aniline. J Molec Struct. 2016;1108:341–346.

[4] Szafran M, Komasa A, Anioła M, Katrusiak A, Dega-Szafran Z. Structure of the complex of dimethylphenyl betaine with dichloroacetic acid studied by X-ray diffraction, DFT calculations, infrared and Raman spectra. Vibrat Spectrosc. 2016;84:92–100.

[5] Agrawal M, Deval V, Gupta A, Sangala BR, Prabhu SS. Evaluation of structure-reactivity descriptors and biological activity spectra of 4-(6-methoxy-2-naphthyl)-2-butanone using spectroscopic techniques. Spectrochim Acta A. 2016;167:142–156.

[6] Pan H, Wang W, Ma Z, Xu L, Zhang Z, Zhao Y, et al. Structures and spectroscopic properties of three [RuCl(2mqn)$_2$NO] (H2mqn = 2-methyl-8-quinolinol) isomers: an experimental and density functional theoretical study. Polyhedron. 2016;118:61–69.

[7] Watson JD, Crick FHC. Molecular structure of nucleic acids – A structure for deoxyribose nucleic acid. Nature. 1953;171(4356):737–738.

[8] Alcolea Palafox M, Iza N, Gil M. The hydration effect on the uracil frequencies: an experimental and quantum chemical study. J Molec Struct (Theochem). 2002;585(1-3):69–92.

[9] Alcolea Palafox M. Scaling factors for the prediction of the frequencies of the ring modes in benzene derivatives. J Phys Chem A. 1999;103:11366–11377.

[10] Kawai T, Ikegami M, Kawai K, Majima T, Nishimura Y, Arai T. Recognition of substituted cytosine derivatives by the base pairing with guanine connected to pyrene. Chem Phys Lett. 2005;407(1-3):58–62.

[11] Steenken S, Reynisson J. DFT calculations on the deprotonation site of the one-electron oxidised guanine–cytosine base pair. Phys Chem Chem Phys. 2010;12(31):9088–9093.

[12] Tripathi J, Tripathi PK, Singh L, Soni RK, Alcolea Palafox M, Rastogi VK. Tautomeric and electronic properties of biomolecule 2-thiocytosine. Int J Curr Chem. 2010;1(2):119–132.

[13] Alcolea Palafox M, Rathor SK, Rastogi R, Bhat D, Jothy BV, Rastogi VK. Asian J Phys. 2015;24(1):33–59.

[14] Alcolea Palafox M, Rastogi VK. The biomolecule of 5-chlorocytosine: geometry of the six main tautomers in the isolated state by DFT calculations, and interpretation of the FT-IR and FT-Raman spectra in the solid state. Asian Chem Lett. 2015;19(1):1–25.

[15] Raczyńska ED, Sapula M, Zientara-Rytter K, Kolczyńska K, Stępniewski TM, Hallmann M. DFT studies on the favored and rare tautomers of neutral and redox cytosine. Struct Chem. 2016;27:133–143.

[16] Yang ZB, Rodgers MT. Theoretical studies of the unimolecular and bimolecular tautomerization of cytosine. Phys Chem Chem Phys. 2004;6:2749–2757.

[17] Morris SM. The genetic toxicology of 5-fluoropyrimidines and 5-chlorouracil. Mutat Res. 1993;297(1):39–51.

[18] Kang JI, Burdzy A, Liu PF, Sowers LC. Synthesis and characterization of oligonucleotides containing 5-chlorocytosine. Chem Res Toxicol. 2004;17:1236.

[19] Krishnakumar V, Balachandran V. Normal coordinate analysis of 5-fluoro, 5-chloro and 5-bromo-cytosines. Ind J Pure Appl Phys. 2001;39(10):623–627.

[20] Krishnakumar V, Balachandran V. Analysis of vibrational spectra of 5-fluoro, 5-chloro and 5-bromo-cytosines based on density functional theory calculations. Spectrochim Acta A. 2005;61(5):1001–1006.
[21] Piacenza M, Grimme S. Systematic quantum chemical study of DNA-base tautomers. J Comput Chem. 2004;25(1):83–98.
[22] Fogarasi G. Relative stabilities of three low-energy tautomers of cytosine: A coupled cluster electron correlation study. J Phys Chem A. 2002;106:1381–1390.
[23] Alparone A, Millefiori A, Millefiori S. Non-planarity and solvent effects on structural and polarizability properties of cytosine tautomers. Chem Phys. 2005;312(1-3):261–274.
[24] Clowney L, Jain SC, Srinivasan AR, Westbrook J, Olson WK, Berman HM. Geometric parameters in nucleic acids: nitrogenous bases. J Am Chem Soc. 1996;118(3):509–518.
[25] Seminario JM, Politzer P. Modern density functional theory: a tool for chemistry Vol. 2. Amsterdam: Elsevier, 1995.
[26] March NH. Electron density theory of atoms and molecules. London: Acad. Press, 1991.
[27] Alcolea Palafox M, Rastogi VK. Quantum chemical predictions of the vibrational spectra of polyatomic molecules. The uracil molecule and two derivatives. Spectrochim Acta A. 2002;58(3):411–440.
[28] Becke AD. Density-functional thermochemistry. 3. The role of exact exchange. J Chem Phys. 1993;98(7):5648–5652.
[29] Lee C, Yang W, Parr RG. Development of the Colle–Salvetti correlation-energy formula into a functional of the electron-density. Phys Rev. 1988;B37(2):785–789.
[30] Frisch MJ, Trucks GW, Schlegel HB, Scuseria GE, Robb MA, Cheeseman JR, et al. Gaussian 09, Revision D.01. Wallingford CT: Gaussian, Inc., 2009.
[31] Scott AP, Radom L. Harmonic vibrational frequencies: An evaluation of Hartree–Fock, Møller–Plesset, quadratic configuration interaction, density functional theory, and semi-empirical scale factors. J Phys Chem. 1996;100(41):16502–16513.
[32] Pulay P, Saebo S, Malagoli M, Baker J. Accuracy and efficiency of atomic basis set methods versus plane wave calculations with ultrasoft pseudopotentials for DNA base molecules. J Comput Chem. 2005;26:599–605.
[33] Radchenko ED, Plokhitnichenko AM, Ivanov AY, Sheina GG, Blagoy YP. Ketoenole tautomeria of guanine and isocytosine molecules. Biofizika. 1986;31(3):373–381.
[34] Nowak MJ, Lapinski L, Fulara J. Matrix isolation studies of cytosine: the separation of the infrared spectra of cytosine tautomers. Spectrochim Acta A. 1989;45(2):229–242.
[35] Szczesniak M, Szczepaniak K, Kwiatkowski JS, KuBulat K, Person WB. Matrix isolation infrared studies of nucleic acid constituents. 5. Experimental matrix-isolation and theoretical ab initio SCF molecular orbital studies of the infrared spectra of cytosine monomers. J Am Chem Soc. 1988;110(25):8319–8330.
[36] Radchenko ED, Sheina GG, Smorygo NA, Blagoi YP. Experimental and theoretical studies of molecular structure features of cytosine. J Mol Struct. 1984;116(3-4):387–396.
[37] Aamouche A, Ghomi M, Coulombeau C, Jobic H, Grajcar L, Baron MH, et al. Neutron inelastic scattering, optical spectroscopies and scaled quantum mechanical force fields for analyzing the vibrational dynamics of pyrimidine nucleic acid bases. 1. Uracil. J Phys Chem. 1996;100(13):5224–5234.
[38] Szczepaniak K, Person WB, Leszczynski J, Kwiatkowski JS. Matrix isolation and DFT quantum mechanical studies of vibrational spectra of uracil and its methylated derivatives. Polish J Chem. 1998;72(2):402–420.
[39] Chandra AK, Nguyen MT, Zeegers-Huyskens T. Theoretical study of the interaction between thymine and water. Protonation and deprotonation enthalpies and comparison with uracil. J Phys Chem A. 1998;102:6010–6016.

[40] Harsányi L, Császár P, Császár A, Boggs JE. Interpretation of the vibrational spectra of matrix-isolated uracil from scaled ab initio quantum mechanical force fields. Int J Quantum Chem. 1986;29(4):799–815.
[41] Ivanov AY, Plokhotnichenko AM, Radchenko ED, Sheina GG, Blagoi YP. FTIR spectroscopy of uracil derivatives isolated in Kr, Ar and Ne matrices: matrix effect and fermi resonance. J Mol Struct. 1995;372(2-3):91–100.
[42] Colarusso P, Zhang K, Guo B, Bernath PF. The infrared spectra of uracil, thymine, and adenine in the gas phase. Chem Phys Lett. 1997;269(1-3):39–48.
[43] Lés A, Adamowicz L, Nowak MJ, Lapinski L. The infrared spectra of matrix isolated uracil and thymine: an assignment based on new theoretical calculations. Spectrochim Acta A. 1992;48(10):1385–1395.
[44] Császár P, Harsányi L, Boggs JE. Vibrational frequencies and assignments for some isotopomers of uracil using a scaled ab initio force field. Int J Quantum Chem. 1988;33(1):1–17.
[45] Rastogi VK, Singh C, Jain V, Alcolea Palafox M. FTIR and FT-Raman spectra of 5-methyluracil (thymine). J Raman Spectrosc. 2000;31(11):1005–1012.
[46] Rastogi VK, Jain V, Yadav RA, Singh C, Alcolea Palafox M. Fourier transform Raman spectrum and ab initio and density functional computations of the vibrational spectrum, molecular geometry, atomic charges and some molecular properties of the anticarcinogenic drug 5-fluorouracil. J Raman Spectrosc. 2000;31(7):595–603.
[47] Ferro D, Bencivenni L, Teghil R, Mastromarino R. Vapour pressures and sublimation enthalpies of thymine and cytosine. Thermochim Acta. 1980;42(1):75–83.
[48] Florian J, Baumruk V, Leszczynski J. IR and Raman spectra, tautomeric stabilities, and scaled quantum mechanical force fields of protonated cytosine. J Phys Chem. 1996;100(13):5578–5589.
[49] Aamouche A, Ghomi M, Grajcar L, Baron MH, Romain F, Baumruk V, et al. Neutron inelastic scattering, optical spectroscopies and scaled quantum mechanical force fields for analyzing the vibrational dynamics of pyrimidine nucleic acid bases: 3. Cytosine. J Phys Chem A. 1997;101:10063–10074.
[50] Ten GN, Baranov VI. Analysis of electronic-vibrational spectra of uracil, thymine, and cytosine. Opt Spectrosc. 2004;97(2):195–203.
[51] Gould IR, Vincent MA, Hillier I, Lapinski L, Nowak MJ. A new theoretical prediction of the infrared spectra of cytosine tautomers. Spectrochim Acta A. 1992;48(6):811–818.
[52] Jaworski A, Szczesniak M, Szczepaniak K, Kubulat K, Person WB. Infrared spectra and tautomerism of 5-fluorocytosine, 5-bromocytosine and 5-iodocytosine. Matrix isolation and theoretical AB initio studies. J Mol Struct. 1990;223:63–92.
[53] Barker DL, Marsh RE. Crystal structure of cytosine. Acta Crystallogr. 1964;17(12):1581.
[54] Alcolea Palafox M, Rastogi VK, Kumar S, Joe H. The biomolecule of 5-bromocytosine: FT-IR and FT-Raman spectra and DFT calculations. Identification of the tautomers in the isolated state and simulation the spectra in the solid state. Spectrochim Acta A. 2013;111:104–122.
[55] Alcolea Palafox M. Scaling factors for the prediction of vibrational spectra. I. Benzene molecule. Int J Quantum Chem. 2000;77:661–684.

K. Gbayo, C. Isanbor, K. Lobb and O. Oloba-Whenu

6 Mechanism of nucleophilic substitution reactions of 4-(4'-nitro)phenylnitrobenzofurazan ether with aniline in acetonitrile

Abstract: Rate constants and activation parameters obtained for the nucleophilic aromatic substitution reactions (S_NAr) of 4-substitutedphenoxy-7-nitrobenzoxadiazole (1) with aniline in acetonitrile at varying temperature using Nuclear Magnetic Resonance (NMR) techniques were reported. These results were compared with the theoretical study which identifies transformations and intermediates using Density Functional Theory (DFT).

Keywords: Nucleophilic aromatic substitution, NMR, DFT

6.1 Introduction

The 7-nitrobenzofurazan derivatives are a product of the nucleophilic aromatic substitution (S_NAr) of the activated chlorine atom on 4-chloro-7-nitrobenzofurazan (4-chloro-7-nitrobenzo-2-oxa-1,3-diazole, NBD chloride, NBD-Cl) **1** by primary or secondary amines, thiols and phenoxide tyrosine anions [1–7]. Hence, nitro-2,1,3-benzoxadiazoles, and related oxide derivatives (commonly referred to as nitrobenzofurazans and nitrobenzofuroxans, respectively) attracted much interest as typical 10 π electrons heteroaromatic substrates possessing an extremely high electrophilic character. The electrophilic character of nitro-2,1,3-benzoxadiazoles derivatives is particularly remarkable in the case of halo-nitrobenzofurazans that easily undergo S_NAr reactions, leading to several analytical and biochemical applications [8]. Some 4-aryloxy-7-nitrobenzofurazan derivatives have biologically active structural fragments, allowing their use in pharmacokinetics and pharmacodynamics. Some have also been reported to possess antileukemic and immunosuppressive activity [9].

The 4-chloro-7-nitrobenzofurazan **1** is a compound with variety of uses ranging from fluorescently labelling amino acids for diagnostic purposes [10, 11] through to the creation of highly sensitive fluorescent/colometric probes for use in analysis. Its use as a building block to generate much larger fluorescent systems justifies studies in the mechanism by which it reacts with nucleophilic molecules. Aromatic nucleophilic substitution of NBD-Cl has drawn attention in the literature and displays a range of reactivity. For instance, susceptibility to attack at position 5 by carbanions has been characterized kinetically [12]. However, the reactivity of NBD-Cl at position 7

https://doi.org/10.1515/9783110467215-006

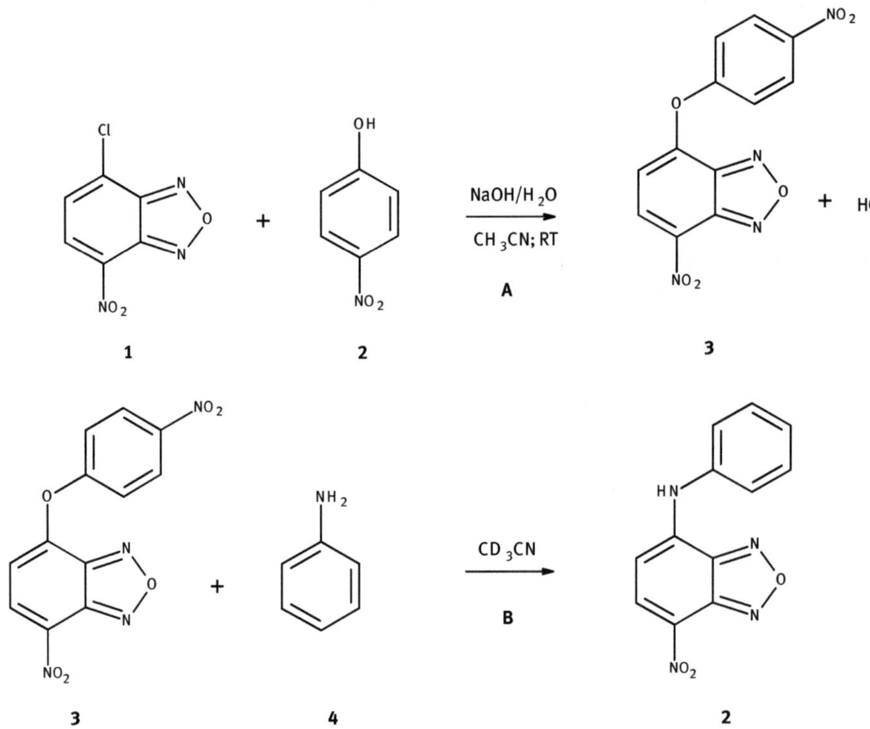

Figure 6.1: Synthesis **A** of the nitrophenoxy adduct, and its aminolysis **B**.

is widely documented and the reactions at this position with indoles [13], indolizines [14] and series of substituted phenols [15] are good examples of this.

In this present work, the nitrophenoxy derivative **3** is first synthesized by reacting NBD-Cl **1** with 4-nitrophenol **2** (Figure 6.1). Activation parameters for the reaction of this derivative, 4-(4'-nitro)phenoxy-7-nitrobenzofurazan **3**, with aniline **4** (Figure 6.1) have been determined by performing this reaction at a variety of temperatures.

6.2 Results and discussion

The mechanism for S_NAr attack by the aniline is expected to follow the mechanism in Figure 6.2 [15]. A σ-complex is initially formed; subsequent to this, a proton transfer and loss of the leaving group results in the reintroduction of the aromaticity.

In CDC_3N, the reaction between 4-(4'-nitro)phenoxy-7-nitrobenzofurazan **3** and aniline **4** is spontaneous at room temperature. Figure 6.3 illustrates the stacked 600 MHz ^{1}H Nuclear Magnetic Resonance (NMR) spectra obtained at time intervals at 328 K for this process. From this, the disappearance of the signal at 8.40 and 8.52 ppm from the phenoxy substrate disappears, while the aniline derivative forms as seen by

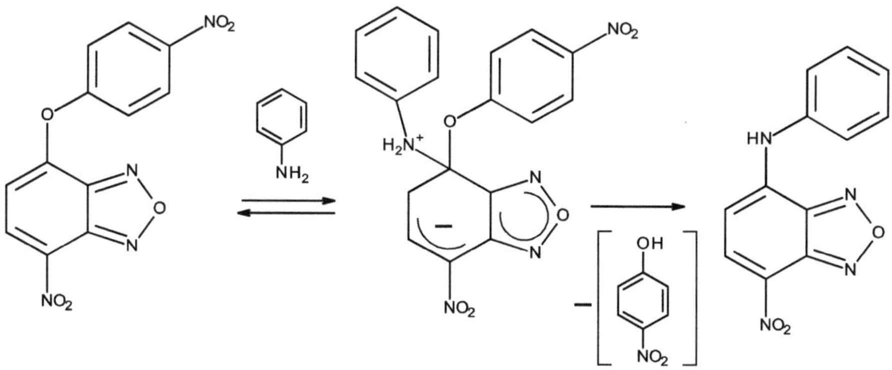

Figure 6.2: S_NAr mechanism forming the σ-complex intermediate, followed by loss of the leaving group.

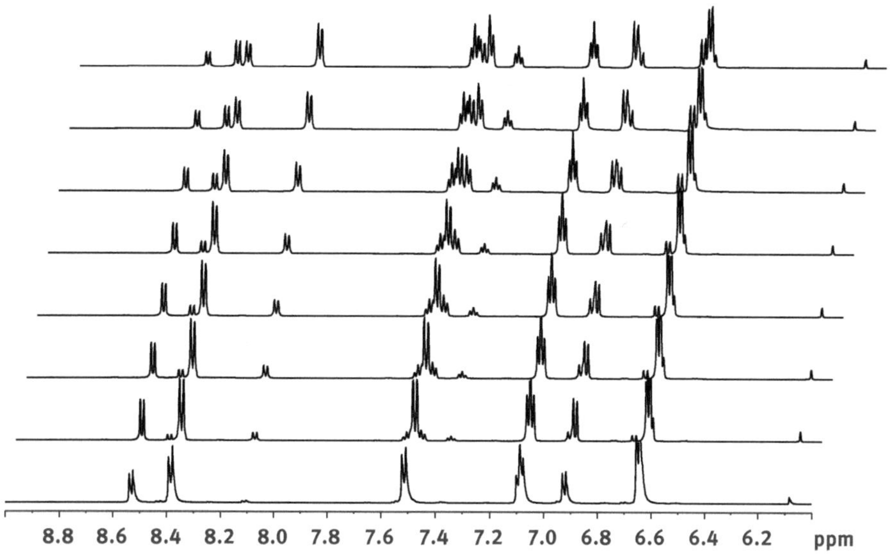

Figure 6.3: Stacked plot of the reaction of NBD-ether and aniline in CD_3CN at 328 K.

the appearance of the signal at 7.28 and 8.10 ppm. Signals from the σ-complex intermediate were not observed in the NMR spectra.

The trimethoxybenzene (TMB) signal at 6.13 ppm remains constant throughout the reaction as expected. The kinetics data were acquired for a range of temperatures (296 K–328 K) and were automatically integrated. The ^{1}H-NMR integral data were used to determine the rate of formation of product and the rate of consumption of the reactants. After conversion of the integrals to the concentrations of the various species in the reaction mixture, the graphs of these concentrations versus time were plotted against a theoretical concentration curve based on a second-order rate law. The plots of experimental versus theoretical concentrations are illustrated in Figure 6.4.

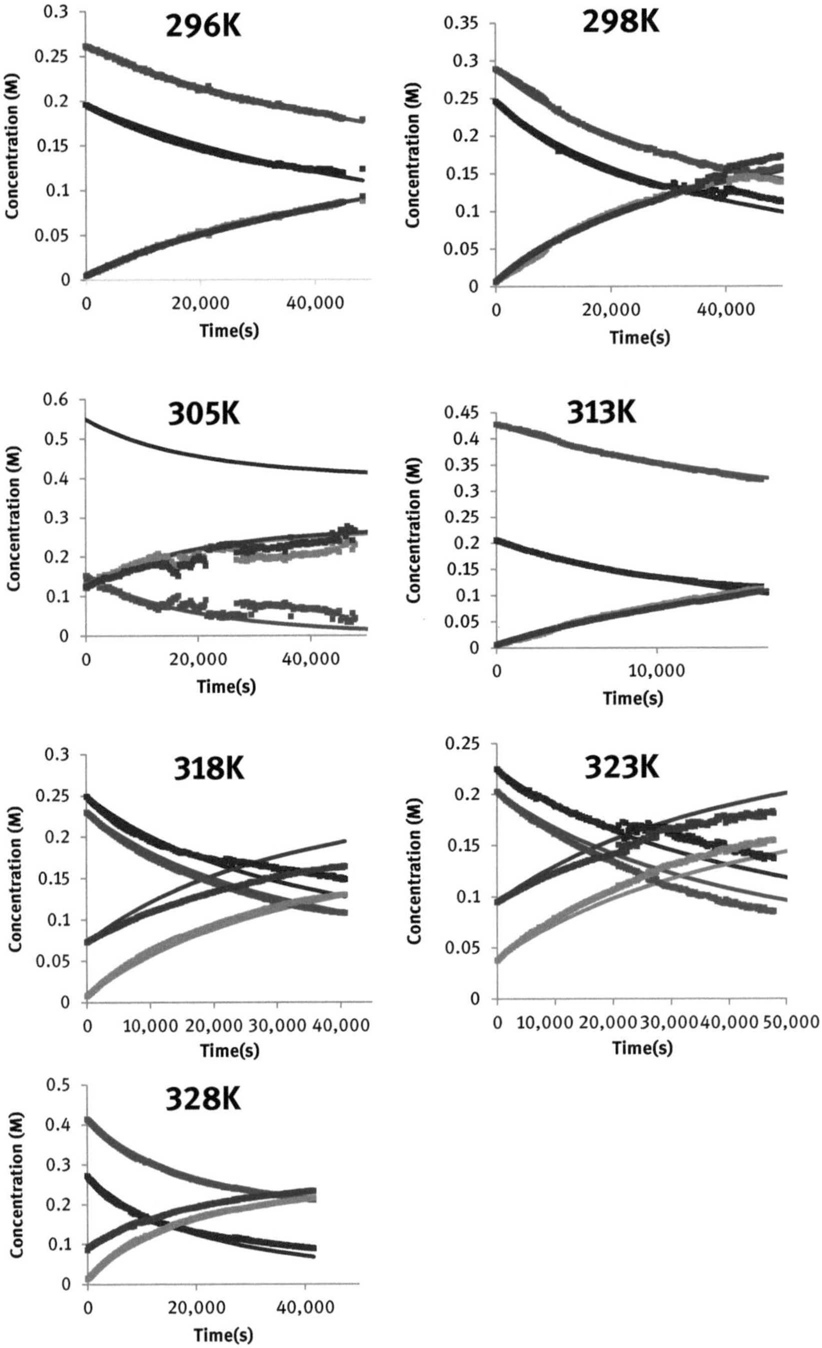

Figure 6.4: Theoretical and experimental superimposed graph of concentration versus time for the reaction of NBD ether with aniline at various temperatures.

Product **5** formation and decrease in the concentration of NBD ether **3** as well as the aniline **4** was evident from the ^{1}H-NMR integral (Figure 6.3). The shapes of the curves for the reactants and products of the theoretical concentration versus time graphs for the proposed model plotted were close to those of the experimental kinetic data (Figure 6.4). Rate constants determined showed a consistency with second-order kinetics. Figure 6.5 shows the Eyring plot for this process with $\Delta G^{\neq} = 23.1 \pm 1.9$ kcal.mol^{-1}, $\Delta H^{\neq} = 2.27 \pm 0.95$ kcal.mol^{-1} and $\Delta S^{\neq} = -69.8 \pm 3.1$ Cal.mol^{-1}.K^{-1}.

The large negative $\Delta S^{\neq}$ value and small positive $\Delta H^{\neq}$ value suggests that the reaction proceeds through a typical rate-determining nucleophilic attack.

In a theoretical study, transition states and intermediates (Figure 6.6) were optimized for the reaction of aniline **4** with 4-(4'-nitro)phenoxy-7-nitrobenzofurazan **3** at the M06-2X/6-31+G(d,p) level [16], and the potential energy surface passing through the σ-complex intermediate is presented in Figure 6.7, with the values reported in Table 6.1. Under these conditions, the activation free energy for the rate-determining initial step was determined to be of the order of 50 kcal/mol. This is of the same order of magnitude as the experimental data, but a more accurate correlation might be obtained by investigation of the microscopic rate constants, or from MP2 or CCSD calculations. The σ-complex has been identified. Proton transfer in this intermediate is responsible for the final transition state and the formation of the final product.

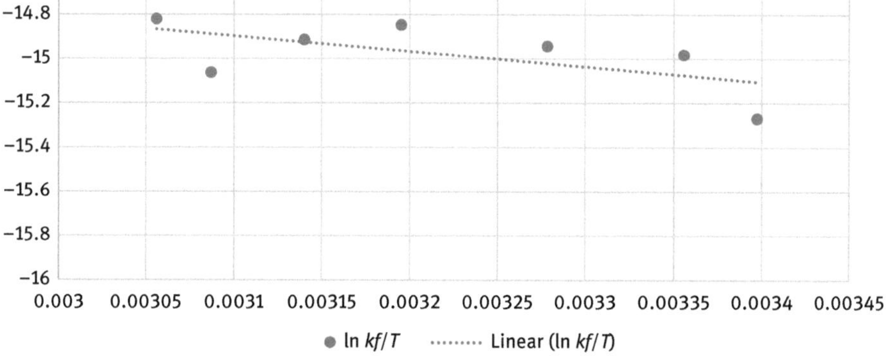

Figure 6.5: Eyring plot for the process.

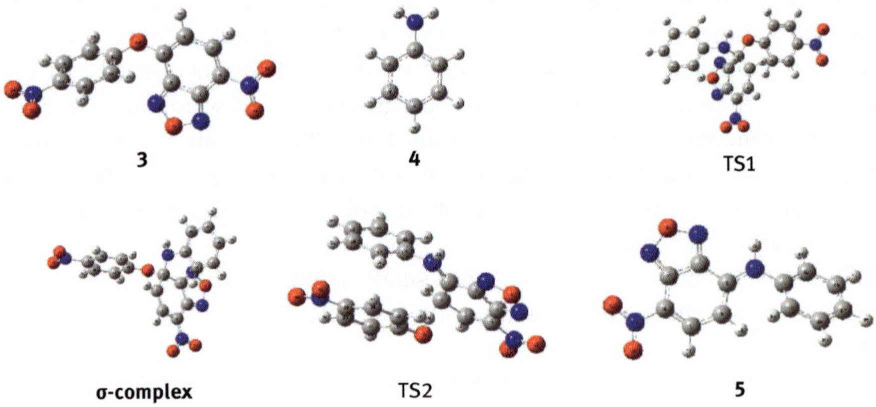

Figure 6.6: Optimized structures of molecules **3**, **4**, TS1, the σ-complex intermediate, TS2 and **5** obtained at the DFT/M06-2x/6-31G+(d,p) calculation in acetonitrile.

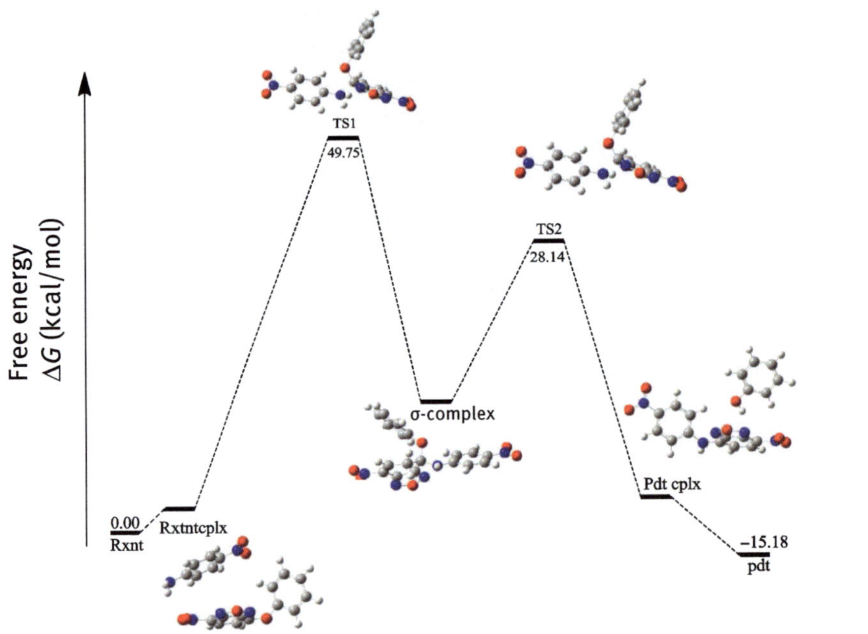

Figure 6.7: Potential energy surface for the transformation of 4-(4'-nitro)phenoxy-7-nitrobenzofurazan **3** to the adduct **5**.

6.3 Conclusion

The kinetic studies of the reaction of 4-(4'-nitro)phenyl-7-nitrobenzofurazan ether with aniline has been carried out in acetonitrile at a variety of

Table 6.1: Thermodynamic and kinetic parameters for studied reaction calculated using DFT/M06-2x/6-31+G(d,p).

	In vacuo			In acetonitrile		
	TS1	**TS2**	**5**	**TS1**	**TS2**	**5**
ΔG^0 (kcal.mol^{-1})	51.42	34.68	−13.25	49.75	28.14	−15.8
ΔH^0 (kcal.mol^{-1})	37.83	21.55	−13.49	35.65	16.10	−15.92
ΔS^0 (cal.mol^{-1}.K^{-1})	−45.60	−44.06	−0.81	−47.31	−40.40	−2.48

temperatures. These results are consistent with a S_NAr-Ad.E reaction with a rate-limiting nucleophilic attack and formation of the intermediate σ complex followed by fast expulsion of the phenoxy leaving group proposed for the anilinolysis of 4-nitrophenyl-7-nitrobenzofurazan ether in acetonitrile. A potential energy surface has been calculated at the M06-2X/6-31+G(d,p) level in support of this mechanism.

6.4 Experimental section

Sodium hydroxide (0.02 g, 0.5 mmol) was dissolved in water (1 cm^3) and 4-nitrophenol (0.0695 g, 0.5 mmol) was dissolved in solution. 7-Chloro-4-nitrobenzofurazan (0.0998 g, 0.5 mmol) was dissolved in acetonitrile (2 cm^3) and added to the phenolate solution while stirring for 1 h. About 10 ml of distilled water was added to precipitate the ether which was then filtered and recrystallized from ethanol.

Kinetics: The progress of the reaction between 4-nitrophenylnitrobenzofurazan ether and aniline to form the corresponding nitrobenzofurazan amine derivative was monitored using the ^{1}H-NMR integral ratios of selected structure-specific signals. The integral data were generated as text and opened in MS-Excel. Trimethoxybenzene (TMB) was used as the internal standard. Kinetic experiments were carried out in CD_3CN at 296 K, 298 K, 305 K, 313 K, 318 k, 323 k and 328 k.

Calculations: All density functional theory (DFT) calculations were performed using the Gaussian 09 package [17]. Molecular geometries were optimized using DFT M06-2X method and 6-31+G(d,p). Vibrational frequencies of all optimized structures were calculated to inspect the nature of the stationary points (minimum or transition structure), and transition state structures are associated with only one imaginary frequency.

Funding: This work is based on the research supported in part by the National Research Foundation of South Africa (Unique Grant No. 94141).

References

[1] Uchiyama Seiichi, Santa Tomofumi, Imai Kazuhiro. Study on the fluorescent 'on–off' properties of benzofurazan compounds bearing an aromatic substituent group and design of fluorescent 'on–off' derivatization reagents. The Analyst. 2000;125(10):1839–1845. DOI:10.1039/B005217P.

[2] Onoda Maki, Uchiyama Seiichi, Santa Tomofumi, Imai Kazuhiro. The effects of spacer length on the fluorescence quantum yields of the benzofurazan compounds bearing a donor-acceptor system. Luminescence. 2002 1;17(1):11–14. DOI:10.1002/bio.670.

[3] Bem M, Caproiu MT, Vasilescu M, Tudose M, Socoteanu R, Nicolae A, Constantinescu T, Banciu M. Synthesis of new fluorescent derivatives of 1,7,10,10-tetraoxa-4,13-diazaclyclooctadecane (Kryptofix K22). Revue Roumaine de Chimie. 2003;48:709.

[4] Lakshmi C., Hanshaw Roger G., Smith Bradley D. Fluorophore-linked zinc(II)dipicolylamine coordination complexes as sensors for phosphatidylserine-containing membranes. Tetrahedron. 2004 11;60(49):11307–11315. DOI:10.1016/j.tet.2004.08.052.

[5] Bem M, Badea F, Draghici C, Caproiu MT, Vasilescu M, Voicescu M. Synthesis and fluorescent properties of new derivatives of 4-amino-7-nitrobenzofurazan. Arkivoc. 2007 6 5;2007(13):87–104. DOI:10.3998/ark.5550190.0008.d12.

[6] Uchiyama Seiichi, Santa Tomofumi, Okiyama Natsuko, Fukushima Takeshi, Imai Kazuhiro. Fluorogenic and fluorescent labeling reagents with a benzofurazan skeleton. Biomedical Chromatography. 2001;15(5):295–318. DOI:10.1002/bmc.75.

[7] Onoda Maki, Uchiyama Seiichi, Endo Atsushi, Tokuyama Hidetoshi, Santa Tomofumi, Imai Kazuhiro. First Fluorescent Photoinduced Electron Transfer (PET) Reagent for Hydroperoxides. Organic Letters. 2003 5;5(9):1459–1461. DOI:10.1021/ol0342150.

[8] Mateeva Nelly N., Deiab Shihab D., Archibong Edikan E., Jackson Mercedes, Mochona Bereket, Gangapuram Madhavi, Redda Kinfe K. N-(4-amino-7-nitrobenzaoxa-1,3-diazole)-substituted aza crown ethers: complexation with alkali, alkaline earth metal ions and ammonium. Journal of Inclusion Phenomena and Macrocyclic Chemistry. 2010 5 7;68(3-4):305–312. DOI:10.1007/s10847-010-9788-2.

[9] Ghosh P, Whitehouse M. Potential Antileukemic and Immunosuppressive Drugs. II. Further Studies with Benzo-2,1,3-oxadiazoles (Benzofurazans) and Their N-Oxides (Benzofuroxans). Journal of Medicinal Chemistry. 1969 5;12(3):505–507. DOI:10.1021/jm00303a606.

[10] Swiecicki Jean-Marie, Di Pisa Margherita, Burlina Fabienne, Lécorché Pascaline, Mansuy Christelle, Chassaing Gérard, Lavielle Solange, et al. Accumulation of cell-penetrating peptides in large unilamellar vesicles: A straightforward screening assay for investigating the internalization mechanism. Biopolymers. 2015 9;104(5):533–543. DOI:10.1002/bip.22652.

[11] Sagirli Olcay, Toker Sıdıka Erturk, Önal Armağan. Development of sensitive spectrofluorimetric and spectrophotometric methods for the determination of duloxetine in capsule and spiked human plasma. Luminescence. 2014 3 12;29(8):1014–1018. DOI:10.1002/bio.2652.

[12] ASGHAR BASIM H. Reactions of Substituted Phenylnitromethane Carbanions with Aromatic Nitro Compounds in Methanol: Carbanion Reactivity, Kinetic, and Equilibrium Studies. International Journal of Chemical Kinetics. 2014 6 26;46(8):477–488. DOI:10.1002/kin.20864.

[13] Rodriguez-Dafonte Pedro, Terrier Francois, Lakhdar Sami, Kurbatov Sergei, Goumont Regis. Carbon Nucleophilicities of Indoles in SNAr Substitutions of Superelectrophilic

7-Chloro-4,6-dinitrobenzofuroxan and -benzofurazan. The Journal of Organic Chemistry. 2009 5;74(9):3305–3315. DOI:10.1021/jo900076r.
[14] Tatarov Artem, Kurbatov Serguey, Borodkin Gennady, Goumont Régis, Terrier François. SEAr–SNAr couplings of indolizines and related pyrrole derivatives with superelectrophilic nitrobenzoxadiazoles. Tetrahedron. 2010 1;66(4):995–1006. DOI:10.1016/j.tet.2009.11.071.
[15] Merouani Hafida, Mokhtari Malika, Ouddai Nadia. Réactions de nitrobenzofurazanes avec des phénols para substitués. Étude cinétique et mécanisme. Comptes Rendus Chimie. 2009 6;12(6-7):816–823. DOI:10.1016/j.crci.2008.10.006.
[16] Zhao Yan, Truhlar Donald G. The M06 suite of density functionals for main group thermochemistry, thermochemical kinetics, noncovalent interactions, excited states, and transition elements: two new functionals and systematic testing of four M06-class functionals and 12 other functionals. Theoretical Chemistry Accounts. 2007 7 12;120(1-3):215–241. DOI:10.1007/s00214-007-0310-x.
[17] Frisch MJ, Trucks GW, Schlegel HB, Scuseria GE, Robb MA, Cheeseman JR, et al. Gaussian 09, Revision E.01. Wallingford CT: Gaussian, Inc., 2016.

Sutapa Biswas Majee and Gopa Roy Biswas

7 Computational methods in preformulation study for pharmaceutical solid dosage forms of therapeutic proteins

Abstract: Design and delivery of protein-based biopharmaceuticals needs detailed planning and strict monitoring of intermediate processing steps, storage conditions and container-closure system to ensure a stable, elegant and biopharmaceutically acceptable dosage form. Selection of manufacturing process variables and conditions along with packaging specifications can be achieved through properly designed preformulation study protocol for the formulation. Thermodynamic stability and biological activity of therapeutic proteins depend on folding–unfolding and three-dimensional packing dynamics of amino acid network in the protein molecule. Lack of favourable environment may cause protein aggregation with loss in activity and even fatal immunological reaction. Although lyophilization can enhance the stability of protein-based formulations in the solid state, it can induce protein unfolding leading to thermodynamic instability. Formulation stabilizers such as preservatives can also result in aggregation of therapeutic proteins. Modern instrumental techniques in conjunction with computational tools enable rapid and accurate prediction of amino acid sequence, thermodynamic parameters associated with protein folding and detection of aggregation "hot-spots." Globular proteins pose a challenge during investigations on their aggregation propensity. Biobetter therapeutic monoclonal antibodies with enhanced stability, solubility and reduced immunogenic potential can be designed through mutation of aggregation-prone zones. The objective of the present review article is to focus on the various analytical methods and computational approaches used in the study of thermodynamic stability and aggregation tendency of therapeutic proteins, with an aim to develop optimal and marketable formulation. Knowledge of protein dynamics through application of computational tools will provide the essential inputs and relevant information for successful and meaningful completion of preformulation studies on solid dosage forms of therapeutic proteins.

Keywords: preformulation, therapeutic proteins, protein aggregation

7.1 Introduction

Formulation development requires a balance between physical and chemical properties to achieve a robust formulation with the desired physical and

https://doi.org/10.1515/9783110467215-007

bioequivalent results. Stages of formulation development include description of formulation platforms for preclinical studies, development of preclinical formulations, pre-formulation studies of active pharmaceutical ingredient (API) to support drug–product development and finally formulation development. Preformulation is a branch of pharmaceutical sciences that utilizes biopharmaceutical principles in the determination of physicochemical properties of a drug substance. Preformulation testing encompasses all studies enacted on a new drug compound in order to develop a stable, elegant and biopharmaceutically suitable dosage form and successful commercial product. Goals of preformulation studies are to choose the correct form of API, to evaluate its physical properties and generate a thorough understanding of the material's stability under various conditions, thereby leading to the optimal, marketable drug delivery system. Preformulation study is therefore considered an exploratory tool initiated early in the development of any pharmaceutical dosage form [1].

7.2 Challenges to formulation development of therapeutic proteins

Significant strides in the field of biotechnology have opened up possibilities for large-scale production of therapeutic proteins which have immense therapeutic potential in the treatment of a range of diseases, starting from rheumatoid arthritis to cancers. Different therapeutic proteins available in the market include monoclonal antibodies, erythropoietins, interferons, growth factors, insulin, interleukins, tissue plasminogen activator, blood clotting factors and replacement enzymes. Monoclonal antibodies are highly preferable owing to their target specificity. Worldwide sales of monoclonal antibodies are expected to reach US $125 billion by 2020 [2]. However, complex nature of these protein macromolecules and their marginal stability even in solid state act as barriers to successful formulation development of therapeutic proteins. Other major problems associated with therapeutic proteins are their exogenous sources, variations in processing condition of the same molecule from one manufacturer to the other, inter-batch differences in the conditions and finally heterogeneity. Since the processing involves numerous steps, minor modification in a single step can be detrimental to the stability of the protein molecule [3,4]. Preformulation studies therefore form an integral part of protein-based product development which will accelerate clinical manufacturing and investigational new drug filing process. A thorough understanding of physicochemical characteristics of the protein molecule is essential towards setting up the preformulation study protocol. Stability of protein molecules is affected by manufacturing variables, sterilization conditions and container-closure system. The two major issues that are of primary concern during development of solid dosage forms for therapeutic proteins include aggregation propensity at various stages of manufacturing and possible immunogenicity with fatal outcomes [5].

Assessment of physical stability of a protein-based therapeutic is a challenging task as it is linked to the primary sequence of amino acids in the protein. Secondary structure of the protein or its conformation results from intra- and intermolecular hydrogen bonds resulting in the formation of a typical geometric shape. The conformations that usually occur are α–helix, β–sheets, turns, bends and irregular random coil [6, 7]. Stability and biological activity of proteins are closely linked to their global flexibility, fluctuations and other dynamic processes. Flexibility which is essential for its activity poses a challenge to the manufacturers as it compromises on the stability. Although the unfolded state of the protein is unstable, the folded state is only marginally stable and any alteration in manufacturing conditions or in the neighbouring environment may cause various types of instability problems for the protein molecule such as aggregation, degradation by oxidation, hydrolysis, deamidation and lastly, inactivation. Thermodynamic stability of proteins is thus a complex balance between folded and unfolded conformations. Protein stability is greatly affected by those factors which disturb the delicate balance between the stabilizing and destabilizing forces. Protein dynamics should be studied in detail in order to develop and optimize manufacturing conditions for protein-based drug delivery systems. During manufacture and storage, proteins are vulnerable towards aggregation, leading to degradation, denaturation and resultant loss in biological activity. Rational design of protein-based biopharmaceuticals and self-assembled protein-inspired supramolecular aggregates requires a thorough understanding of the protein structure, thermodynamic events and factors responsible for switching of conformation. Information about the thermodynamics of the process of unfolding enables redesigning of proteins, enzymes and antibodies with better solubility and stability through modification of amino acid sequence [8].

7.3 Aggregation of therapeutic proteins

Protein aggregation is a natural phenomenon because it results in the formation of non-covalent bonds which are very similar to those present in native, stabilized structure [9]. Aggregation-induced degradation with antibodies may evoke fatal immunological reaction. Elevated temperature during storage or mechanical vibration during transportation can contribute to aggregation [10]. Various other physical stress factors have been found equally responsible to cause aggregation and include pH, ionic strength, presence of metal ions and surface adsorption [11]. Although stability of protein-based dosage forms can be enhanced by the process of lyophilization or freeze-drying, the dried powder needs to be reconstituted with suitable vehicle prior to administration. However, aggregation may also occur in the lyophilized proteins. Lyophilization process variables and composition of the solid determine the extent of aggregation which is greatly affected by the nature of the protein [12].

Stringent processing conditions during lyophilization may result in thermodynamic instability and reversible unfolding of the protein. If no irreversible damages occur to the protein during storage or reconstitution, then refolding takes place as soon as product is reconstituted, and there is no negative impact on the pharmaceutical stability of the product [13]. Aggregation in proteins can also be induced by the antimicrobial preservatives added to multiple-dose protein formulations. Therefore, suitable biophysical computational tools and instrumental analytical techniques are essential for prediction of aggregation propensity and profiling and for studying the factors affecting alteration in conformation [14].

Local interactions (e. g. hydrogen bonding, steric interactions) are responsible for the formation of α- and β-sheets, loop structure which govern packing dynamics of the molecule to a specific 3D-structure and thus dictate the specific biological function exhibited by the protein. These interactions are constituted by networks of amino acid residues. These networks are assumed to represent a subset of all potential interactions of residues. Literature surveys have revealed the availability of different structural and computational methods to provide insight into amino acid networks. The methods which can provide valuable information even in the absence of any data on protein structure include multiple sequence alignments (MSAs) and algorithms like statistical coupling analysis (SCA), mutual information (MI), McLachlan-based substitution correlation (McBASC) and observed minus expected square (OMES). These computational methods will guide the scientists in the design of engineered proteins with enhanced biological function or biobetters [15]. Physicochemical properties of amino acids, presence of aromatic side chains and charged residues are determinants for aggregation tendency of protein molecules [16]. There are various approaches for correlating solid-state protein stability with inherent protein characteristics such as glass transition temperature, moisture content and free volume measurement with the help of instrumental techniques relying on bulk properties or population averaged properties. However, meaningful correlation cannot be established always because protein instability is highly dependent on the nature of the functional groups [12].

7.3.1 Instrumental methods of analysis

Different instrumental methods can be utilized for the determination of protein structure and quantification of thermodynamic parameters controlling the mechanisms of protein folding-unfolding. They include differential scanning calorimetry (DSC), X-ray diffraction (XRD) crystallography, circular dichroism (CD) spectroscopy, nuclear magnetic resonance (NMR) spectroscopy and infrared (IR) spectroscopy. DSC is used to measure protein's change in molar heat capacity, enthalpy, entropy and transition midpoint to account for transition of 50 % of protein molecules. DSC study of very dry proteins in solid state indicates that heat of denaturation and heat capacity change on denaturation are very close to the values for solution. This does not mean that proteins in dry state exist in native state but indicates that tertiary structure is present. Transition during unfolding can be reversible, two state and highly cooperative, which can be computed from DSC. The method is capable of estimating the shape of folding free energy surface when fitted to a specific model. Accuracy of prediction depends on the proper selection of model. Model independent estimates of folding barrier height can be obtained with Bayesian probabilistic approach. Ultrafast folding proteins are those where conformational transition occurs in sub-millisecond timescales. Modern analytical tools in conjunction with newer computational methods help in characterization of such ultrafast folding proteins exhibiting non-cooperative folding with no sharp demarcation between thermodynamic states having different energy landscapes. Ultrafast kinetic perturbation method employs laser pulses as triggers to stimulate very rapid shift in folding–unfolding thermodynamic equilibrium under the effect of changes in temperature, pH, pressure or chemical potential. Relaxation rate can be determined by spectroscopic monitoring of the shift and microscopic rates of interconversion between the species can be estimated by analysing with a suitable kinetic model. In molecular dynamic (MD) simulation technique, potential energy of the protein is calculated as a function of atomic coordinates. For the purpose, a simulation box is constructed using all atoms present in the protein molecule together with the neighbouring solvent molecules, numerical integration of Newton's equation of motion is performed where position, force and velocities of each individual atom are defined over a very short time span followed by analysis and establishment of correlation between molecular structure and terabytes of data obtained from simulated trajectories. Atomistic MD simulation helps in deriving mechanistic information. Combining NMR dynamics experiment with DSC also facilitates the characterization of ultrafast folding proteins. These proteins can be used in the design of high-performance biosensors [8, 13, 17–20].

Valuable information about protein structure dynamics can be extracted from XRD studies on the protein crystal. However, traditional crystallographic method removes sharp Bragg reflections or report on correlations in charge density variations and hence fail to produce a clear image of protein motion. Diffractograms from multiple crystal orientations are integrated into a 3D dataset where statistical averaging of the signal is done. To overcome this problem, a new 3D modelling and data extraction technique has been introduced in the investigation of protein crystal structure which produces a sharp image. In this method, previously discarded data from diffuse XRD have been processed using computer modelling [21, 22].

Despite advantages of the new technique, XRD analysis requires a single purified well-ordered crystal of the protein molecule which becomes difficult to obtain most of the times. Moreover, the technique is expensive and time consuming. CD spectrophotometry helps in identification of secondary and tertiary structures on the basis of distinct chiral positioning of the amide chromophores and the local environment of aromatic chromophores. Accuracy level of CD spectral data becomes lower if the secondary structure is rich in β-sheets or comprises a mixture of α-helices and β-sheets [14]. Multidimensional NMR spectroscopy is advantageous, but it is time-consuming when one attempts to obtain high resolution of the structure and it requires large amount of sample. The method is not suitable for proteins with short half-life or low stability at room temperature. Fourier-transformed infrared (FTIR) spectroscopy generates quick response and can provide high-resolution results for small soluble proteins, lyophilized proteins as well as large membrane proteins and low sample weight is needed for the characterization. Moreover, FTIR spectroscopy helps in the selection of suitable cryoprotectant excipients to be used as stabilizers during lyophilization. Amide groups in proteins are groups with IR-active vibrations. Change in the position of amide band I accounts for alteration in the secondary structure and is used as an index of different conformations in the protein molecule. However, amide band I position is determined in an aqueous environment and spectral subtraction needs to be done for elimination of water absorption. Therefore, a novel method of drop coat deposition (DCD) for the protein sample preparation, coupled with confocal Raman spectroscopy, has been employed for structure determination of bovine serum albumin and ovalbumin. Peak fitting parameters obtained from ovalbumin can form the basis for the estimation of secondary structure of unknown immunoglobulin-based monoclonal antibodies and fusion drugs. The novel technique can predict secondary structure of any type with high accuracy in a short time and negating the need for spectral processing of raw data [6, 7]. Recently, few studies have been undertaken to determine the secondary structure composition of therapeutic proteins through second-derivative of amide I region of FTIR spectra via principal component analysis, spectral correlation coefficient, and area of overlap. These approaches could compare and detect minor changes in higher-order structure induced by small changes in solution condition such as temperature, pH, salt concentration and type. Another more sensitive technique, known as soft independent modelling of class analogy (SIMCA), performs chemometric analysis to determine the class or group of observations. The technique is very helpful in the case of multiple samples from each condition. Lyophilization has been found to induce varying degree of changes in protein conformation, such as appearance of new bands and disappearance of solution bands. The spectral correlation coefficient values may vary between 0.5 and greater than 0.9. Bandwidths increase and band positions may shift, indicating disorder in configuration. Loss in water due to freeze-drying may result in the formation of β-sheets [13, 23, 24]. Aggregation during freeze-drying can be remarkably prevented by rapid freezing of the entire solution at the same time which will minimize thermal transitions and glass transitions [4]. The low solubility and non-crystalline nature of protein aggregates create hindrance to their high-resolution characterization by XRD and NMR spectroscopy. Low-resolution data obtained from various instrumental methods like fibre diffraction, electron microscopy, hydrogen–deuterium exchange and electron paramagnetic resonance spectroscopy can be used for deciphering the conformation to a limited degree. Computational approaches can extract valuable information from low-resolution data. These tools are validated using MD simulations [25].

7.3.2 Computational approaches in study of aggregation

Computational algorithm methods have been developed where spatial aggregation propensity (SAP) has been employed for prediction of aggregation-prone areas in monoclonal antibodies, utilizing MD simulation approach. Information on the structure and amino acid sequence forms the basis, which ultimately facilitates the identification of exposed hydrophobic patches/motifs or aggregation-prone zones [10, 14, 26]. Several antimicrobial phenolic preservatives (m-cresol, phenol, benzyl alcohol, phenoxyethanol and chlorobutanol) have been found to induce destabilization, protein unfolding and aggregation to a different extent in protein-based multidose preparations [27]. Antimicrobial preservative, benzyl alcohol, has been found to induce aggregation in IFN-α-2a, when investigated by multiple aggregation propensity programs. Benzyl alcohol may perturb the predicted "hot-spots" in the protein molecule, disrupt the contacts between the critical residues, may initiate local unfolding event or may cause swap of domain. The SAP approach is highly accurate although time consuming. However, the expenditure in full antibody atomistic simulation is quite high. There are cheaper and more viable options such as antibody fragment (Fab, Fc) simulations, implicit solvent models or direct computations from a static structure. The last method is very fast but compromises on the level of accuracy. But it is highly suitable in high-throughput screening of therapeutic protein candidates. Identification of the most stable and soluble analogue of mAbs is possible through adoption of SAP. Aggregation "hot-spots" in the IFN-α-2a can be blocked through mutations which can prevent preservative-induced aggregation [28]. Fibroblast growth factor 21 controls hormone levels and may prove highly beneficial in the management of metabolic disorders like diabetes, obesity, cardiovascular disease, etc. However, multiple-dose formulations of the hormone regulator for parenteral delivery contain a phenolic preservative which induces the aggregation of the protein and lower solubility. With an aim to reduce the aggregation tendency without compromising its therapeutic utility, a variant of the molecule has been developed [29]. Similar approach has also been adopted for the design of biobetter therapeutic mAb and becavizumab. Stabilized mAb was produced through single-point mutations of aggregation-prone zones and design of engineered glycosylated protein where the carbohydrate moiety has been used to mask the "hot-spots" [30]. However, difficulties arise during prediction of aggregation in folded globular proteins because aggregation hot-spots are not exposed or they are buried inside the hydrophobic core and may be discontinuous. They may be involved in cooperative non-covalent interactions sustaining secondary and tertiary structure of the protein. Aggregation propensity of globular proteins can be studied by AGGRESCAN 3D Dynamic mode and extracted information can be utilized to engineer protein variants or mutants with increased solubility and stability. It demonstrates higher accuracy in comparison to first-generation sequence-based programs [9].

7.4 Computational tools in assessment of immunogenicity of therapeutic proteins

Therapeutic antibodies, exogenous enzymes, signalling peptides of well-defined structure and function can be immunogenic, if they are of non-human origin. Non-immunogenic or de-immunized variant of the antibody, immunotolerant variant of the enzyme with unaltered binding specificity, affinity and stability can be developed by grafting of key functional residues from an exogenous therapeutic antibody onto human antibody framework. Grafting can be done on a trial-and-error basis or it can be done on a rational basis with a prior knowledge of structure–function relationship. An important obstacle to grafting is the absence of common modular structure in these therapeutic proteins and also lack of homologous human counterpart. Computational methods have been used on a small subpopulation of peptides (T-cell epitopes) for identification of mutations to minimize MHC II binding, responsible for eliciting immune response. Epitope databases like Immune Epitope Database (IEDB) and the proprietary T Cell Epitope Database™ (TCED™) have also been used for the purpose. Improved computational algorithms will target amino acid sequence for prediction of immunogenicity of the entire protein and will ultimately reveal the best structure with reduced immunogenicity potential. Such dynamic programming-based algorithms will enable mutations at the most promiscuous amino acids that are elements of multiple overlapping immunogenic peptides and resulting in the elimination of over six epitopes per mutation [31, 32]. Although several predictive models have been employed till date, the success rate in accurate prediction is not very encouraging. Low success rate is attributed to lack of sufficient knowledge relevant to the underlying mechanisms for immunogenicity of therapeutic proteins [33].

7.5 Conclusion

Owing to their large, complex and conformationally heterogeneous structures, therapeutic proteins are vulnerable to different instability problems induced by various physicochemical and mechanical stress factors during the various processing steps starting from the first process in multi-step manufacturing to administration. Propensity of the molecules towards aggregation needs attention as it can have deleterious effect on biological activity and safety. Therefore, preformulation studies need to be conducted prior to development of stable and efficacious solid dosage forms for the therapeutic proteins. Stability assessment is of primary concern in the preformulation study which utilizes different low-, intermediate- and high-resolution instrumental methods. Application of computational tools to low-resolution data helps in the prediction of secondary and tertiary structure of proteins with high level of accuracy. Data extracted from such analyses can be exploited in the redesigning and engineering of therapeutic proteins or biobetters with improved solubility and stability, which can be easily developed into a stable and effective formulation.

References

[1] Preformulation means characterization. Aug 2004. Available at: Accessed14Aug2016 http://www.biopharminternational.com/preformulation-means-characterization.

[2] Kang J, Lin X, Penera J. Rapid formulation development for monoclonal antibodies. July 2016. Available at: Accessed:9July2016 http://www.bioprocessintl.com/manufacturing/formulation/rapid-formulation-development-for-monoclonal-antibodies/.

[3] Awotwe-Otoo David, Agarabi Cyrus, Keire David, Lee Sau, Raw Andre, Yu Lawrence, Habib Muhammad J., Khan Mansoor A., Shah Rakhi B. Physicochemical Characterization of Complex Drug Substances: Evaluation of Structural Similarities and Differences of Protamine Sulfate from Various Sources. The AAPS Journal. 2012 6 8;14(3):619–626. DOI:10.1208/s12248-012-9375-0.

[4] Vazquez-Rey M, Lang DA. Aggregates in monoclonal antibody manufacturing processes. Biotechnol Bioeng. 2011;108:1494–1508 Validity of computational tools. BioDrugs 2010, 24, 1.

[5] Hovgaard L, Frokjaer S, van de Weert M. Pharmaceutical formulation development of peptides and proteins. USA: CRC Press, 2012.

[6] Combs JD, Gonzalez CU, Wang C. Surface FTIR techniques to analyze the conformation of proteins/peptides in H_2O environment. J Phys Chem Biophys. 2016;6:202. DOI:10.4172/2161-0398.1000202.

[7] Peters J, Luczak A, Ganesh V, Park E, et al. Protein secondary structure determination using drop coat deposition confocal Raman spectroscopy. Spectroscopy. 2016;31:31–39.

[8] Characterisation of protein stability using Differential Scanning Calorimetry. Aug 2015. Available at: Accessed:24Aug2016 http://www.news-medical.net/whitepaper/20150624/Characterization-of-Protein-Stability-Using-Differential-Scanning-Calorimetry.aspx.

[9] Zambrano R, Jamroz M, Szczasiuk A, Pujols J, et al. AGGRESCAN3D (A3D): server for prediction of aggregation properties of protein structures. Nucleic Acids Res. 2015;1. DOI:10.1093/nar/gkv359.

[10] Clark RH, Latypov RF, Imus CD, Carter J, et al. Remediating agitation-induced antibody aggregation by eradicating exposed hydrophobic motifs. mAbs. 2014;6:1540–1550.

[11] Challener CA. Excipient selection for protein stabilization. Pharm Technol. 2015;Supplement(3) s35–s39.

[12] Moorthy BS, Schultz SG, Kim SG, Topp EM. Predicting protein aggregation during storage in lyophilized solids using solid state amide Hydrogen/Deuterium Exchange with Mass Spectrometric Analysis (ssHDX-MS). Mol Pharmaceutics. 2014;11:1869–1879.

[13] Pikal MJ. Mechanisms of protein stabilization during freeze-drying storage: the relative importance of thermodynamic stabilization and glassy state relaxation dynamics. In: Rey L, editors. Freeze-drying/lyophilization of pharmaceutical and biological products. USA: CRC Press, 2016.

[14] Biophysical analysis in support of development of protein pharmaceuticals. 2016. Available at: Accessed:252016 http://basicmedicalkey.com/biophysical-analysis-in-support-of-development-of-protein-pharmaceuticals/.
[15] O'Rourke KF, Gorman SD, Boehr DD. Biophysical and computational methods to analyze amino acid interaction networks in proteins. Comput Struc Biotechnol J. 2016;14:245–251.
[16] Tartaglia GG, Cavalli A, Pellarin R, Caflisch A. Prediction of aggregation rate and aggregation-prone segments in polypeptide sequences. Prot Sci. 2005;14:2723–2734.
[17] Johnson CM. Differential scanning calorimetry as a tool for protein folding and stability. Arch Biochem Biophys. 2013;531:100–109.
[18] Ibarra-Molero B, Naganathan AN, Sanchez-Ruiz JM, Muñoz V. Modern analysis of protein folding by differential scanning calorimetry. Methods Enzymol. 2016;567:281–318.
[19] Farber P, Darmawan H, Sprules T, Mittermaier A. Analyzing protein folding cooperativity by differential scanning calorimetry and NMR spectroscopy. J Am Chem Soc. 2010;132:6214–6222.
[20] Munoz V, Cerminara M. When fast is better: protein folding fundamentals and mechanisms from ultrafast approaches. Biochem J. 2016;473:2545–2559.
[21] Holistic data analysis, modeling poised to transform protein X-ray crystallography. June 2016. Available at: Accessed:29June2016 https://www.sciencedaily.com/releases/2016/03/160329185327.htm.
[22] van Benschoten AH, Liu L, Gonzalez A, Brewster AS, et al. Measuring and modeling diffuse scattering in protein X-ray crystallography. PNAS. 2016;113:4069–4074.
[23] Stockdale G, Murphy BM, D'Antonio J, Manning MC, Al-Azzam W. Comparability of higher order structure in proteins: chemometric analysis of second-derivative amide I Fourier transform infrared spectra. J Pharm Sci. 2015;104:25–33.
[24] D'Antonio J, Murphy BM, Manning MC, Al-Azzam WA. Comparability of protein therapeutics: quantitative comparison of second-derivative amide I infrared spectra. J Pharm Sci. 2012;101:2025–2033.
[25] Xu Y, Xu D, Liang J. Computational methods for protein structure prediction and modeling: volume 1: basic characterization. USA: Springer Science & Business Media, 2007.
[26] Chennamsetty N, Voynoy V, Kayser V, Helk B, et al. Prediction of aggregation prone regions of therapeutic proteins. J Phys Chem B. 2010;114:6614–6624.
[27] Hutchings RL, Singh SM, Cabello-Villegas J, Mallela KMG. Effect of antimicrobial preservatives on partial protein unfolding and aggregation. J Pharm Sci. 2013;102:365–376.
[28] Bis RL, Singh SM, Cabello-Villegas J, Mallela KMG. Role of benzyl alcohol in the unfolding and aggregation of interferon α-2a. J Pharm Sci Pharm Biotechnol 2014. DOI:10.1002/jps.24105.
[29] Kharitonenkov A, Beals JM, Micanovic R, Strifler BA, et al. Rational design of a fibroblast growth factor 21-based clinical candidate, LY2405319. Plos ONE. 2013;8:58575.
[30] Courtois F, Agrawal NJ, Lauer TM, Trout BL. Rational design of therapeutic mAbs against aggregation through protein engineering and incorporation of glycosylation motifs applied to bevacizumab. mAbs. 2016;8:99–112.
[31] Parker AS, Zheng W, Griswold KE, Bailey-Kellogg C. Optimisation algorithms for functional deimmunization of therapeutic proteins. BMC Bioinformatics. 2010;11:180.
[32] Bryson CJ, Jones TD, Baker MP. Prediction of immunogenicity of therapeutic proteins. Bio Drugs. 2010;24(1):314–322.
[33] Brinks V, Weinbuch D, Baker M, et al. Preclinical models used for immunogenicity prediction of therapeutic proteins. Pharm Res. 2013;30:1719.

Prabhat Ranjan, Tanmoy Chakraborty and Ajay Kumar

8 Computational Investigation of Cationic, Anionic and Neutral Ag_2Au_N (N = 1–7) Nanoalloy Clusters

Abstract: The study of bimetallic nanoalloy clusters is of immense importance due to their diverse applications in the field of science and engineering. A deep theoretical insight is required to explain the physico-chemical properties of such compounds. Among such nanoalloy clusters, the compound formed between Ag and Au has received a lot of attention because of their marked electronic, catalytic, optical and magnetic properties. Density Functional Theory (DFT) is one of the most successful approaches of quantum mechanics to study the electronic properties of materials. Conceptual DFT-based descriptors have turned to be indispensable tools for analysing and correlating the experimental properties of compounds. In this report, we have investigated the ground state configurations and physico-chemical properties of $Ag_2Au_N^{\lambda}$ (N = 1–7, λ = ±1, 0) nanoalloy clusters invoking DFT methodology. Our computed data exhibits interesting odd-even oscillation behaviour. A close agreement between experimental and our computed bond length supports our theoretical analysis.

Keywords: bi-metallic nanoalloy, Conceptual Density Functional Theory, global descriptors

8.1 Introduction

In the recent decades, nanotechnology and its applications have deeply integrated into human's daily life due to its large-scale applications. Since last few years, nanotechnology has emerged as a new research dimension in the domain of science and engineering [1]. The classification of nanoparticles has been done in terms of size range, which has at least one dimension in the range of 1 to 100 nm. That particular size range exists between the levels of atomic/molecular and bulk materials [1–5]. Nanoparticles possess unique physico-chemical properties due to existence of a large number of quantum mechanical and electronic effects [2–4]. However, there are still some instances of nonlinear alteration which shows certain physical properties may vary depending on their size, shape and composition [6]. There are a large number of available scientific reports which illustrate the effects on size and geometry to change the electronic, magnetic, optical, mechanical, chemical and other physical properties of nanoparticles [1, 3, 4]. A deep insight into the study of nanoparticles with well-defined size and geometry may open some better alternatives [7]. The nanoparticles

https://doi.org/10.1515/9783110467215-008

due to its numerous applications in the field of nanotechnology, nano-electronics, material science, bio-medicine, catalysis etc. are of great interest [1, 3, 7–10].

Now a days, study of noble metal clusters such as Cu, Ag, and Au has gained a significant due to their marked optical, electronic, catalytic and magnetic properties [11–14, 16–19]. Especially, gold nanoclusters are very much popular and potential candidates for fabrication, bio-physics, and medical applications due to its unique catalytic, optical and electronic properties [20–35]. Study of bimetallic nanoalloy clusters has emerged as an important topic from both theoretical and experimental point of view, since they exhibit new type of characteristics which are distinct from the pure and bulk metallic systems [11]. In recent days, different compositions of nanoalloys are being utilized for advancement of methodologies and characterization techniques [13, 19, 21]. The study based on core shell structure of nano-compounds is very popular because its properties can be altered through proper control of other structural and chemical parameters. Group 11 metal (Cu, Ag and Au) clusters possess filled d orbitals and one unpaired electron in s shell [22]. This peculiar electronic arrangement is responsible to reproduce the exactly similar shell effects, which are experimentally observed in the alkali metal clusters [23–30].

Many experimental and theoretical studies have been reported on Ag-Au nanoalloy clusters. Negishi et al. [36] have studied geometrical structure and electronic states of $Au_nAg_m^-$ ($2 \le n + m \le 4$) clusters using photoelectron spectroscopy technique. Heiles et al. [37] have investigated the composition of 8-atom Au-Ag clusters using GA-DFT approach. Transition from 2D to 3D between Au_6Ag_2 and Au_5Ag_3 clusters are reported. Weis et al. [38] have reported ion mobility measurement and DFT calculations of $Ag_mAu_n^+$, m+n<6 clusters. In that report, they have pointed out that trimeric ions are in generally triangular shape whereas most of the tetrameric clusters are in rhombus shape, with the exception of Ag_3Au^+. Lee et al. [39] have studied structure and electronic properties of Ag, Au and Ag-Au clusters invoking DFT calculations. They have mentioned that in even number clusters are more stable as compared to odd number for neutral state, however in anionic state reverse trend is observed. Geometries, ionization energies and vertical detachment energies for neutral, cationic and anionic bimetallic Ag_mAu_n $3 \le m + n \le 5$ clusters are also reported by Bonacic- Koutecky et al. [40]. Recently, Tafoughalt et al. [41] have reported ground state configurations, binding energy, HOMO–LUMO gap, magnetic moment, ionization energy, electron affinity and chemical hardness of Ag_nAu_m clusters, up to eight atoms. The result indicate that addition of Au atoms in the cluster exhibits in high binding energy and ionization potential. Shayeghi et al. [42] have computed electronic and optical properties of $Ag_nAu_{4-n}^+$ ($n = 1$–3) clusters in terms of DFT method and made a comparative analysis with experimental data. Zhao et al. [43] have reported computational study on structures, stabilities, growth-pattern behaviours and electronic properties of bimetallic M_2Au_n (M = Ag, Cu; $n = 1$–10) clusters. Their result indicates that Ag_2Au_4 and Cu_2Au_2 clusters are the most stable structures. A study of Geometry and electronic properties of Au_mAg_n ($2 \le m + n \le 8$) clusters has been reported by Zhao et al. [44]. Investigation of electronic and photoluminescence properties of $Ag_nAu_{25-n}{}^{2+}$ nanoclusters in terms of time-dependent density functional calculations has been done by Miranda et al. [45]. The result suggests that silver doping improves the fluorescence behaviour. In another study Tafoughalt et al. [46] have also discussed binding energies, second-order difference in energies, dissociation energies and HOMO–LUMO gap of $AgAu_{n-1}$ ($n = 3$–13) clusters. Recently, Shayeghi et al. [47] have reported vibrational spectra of $Au_nAg_m{}^+.Ar_k$ ($n + m = 4,5$; $k = 1$–4) clusters invoking FIR-MPD spectroscopy in terms of Messenger-Atom technique and Density Functional Theory (DFT) methodology. The computational data are in similar line with the FIR-MPD experiments. Chi et al. [48] has proved that gold improves the stability of AgAu cluster. The binding energy of the bimetallic clusters also increases with the increasing number of gold atoms.

DFT is one of the most popular approaches for the quantum mechanical computation of molecular system [49]. Due to its computational friendly behaviour, DFT is widely accepted method to study the many-body systems. In the domain of material science research, particularly in super conductivity of metal-based alloys, structural, electronic, magnetic and optical properties of nano-alloy clusters, quantum fluid dynamics, molecular dynamics, nuclear physics, DFT has gained a huge importance [50]. The study of DFT covers three major domains viz. theoretical, conceptual and computational [51–54]. Conceptual DFT is established as an important approach to study the chemical reactivity of materials [55–57]. The Conceptual DFT is highlighted following Parr's dictum "Accurate calculation is not synonymous with useful interpretation. To calculate a molecule is not to understand it" [58]. We have rigorously applied conceptual density functional-based local and global descriptors to study physico-chemical properties of nano-engineering materials and drug designing process [59–69].

In the present venture, we have investigated bi-metallic nanoalloy clusters of $Ag_2Au_N{}^\lambda$ ($N = 1$–7, $\lambda = \pm1,0$) invoking DFT-based global descriptors. An attempt has been made to correlate the properties of the compounds with their computational counterparts.

8.2 Computational details

A computational investigation on the bi-metallic nanoalloy clusters of $Ag_2Au_N{}^\lambda$ (N = 1–7, $\lambda = \pm1,0$) is performed invoking electronic structure theory. Geometry optimizations have been done using Gaussian 03 [70] within DFT framework. The Local Spin Density Approximation [71] (LSDA) of exchange correlation has been adopted with basis set LanL2dz [72–75]. LSDA functional is popular in computational study of the materials [65, 76–80]. High accuracy of LanL2dz basis set for metallic clusters is already reported [6, 43, 80–83]. In spite of less complexity of LDA functional, it is highly effective in study of solid state physics [71], where accurate phase transitions in solids [72] and liquid metals [73, 74] are predicted, and for lattice crystals, in which 1 % precision are successfully achieved [75].

Restriction on molecular spin is not imposed during geometry optimization. Z-axis is considered as spin polarization axis and Symmetrized Fragment Orbitals (SFOs) are combined with auxiliary Core Functions (CFs) to ensure orthogonalization on the (frozen) Core Orbitals (Cos) [84, 85].

Invoking Koopman's approximation [55], Ionization Energy (I) and Electron Affinity (A) of all the nanoalloys have been computed in terms of following

$$I = -\varepsilon_{HOMO} \tag{8.1}$$

$$A = -\varepsilon_{LUMO} \tag{8.2}$$

Thereafter, using I and A, the conceptual DFT-based descriptors viz. electronegativity (χ), global hardness (η), molecular softness (S) and electrophilicity index (ω) have been computed as:

$$\chi = -\mu = \frac{I+A}{2} \tag{8.3}$$

where μ represents the chemical potential of the system.

$$\eta = \frac{I-A}{2} \tag{8.4}$$

$$S = \frac{1}{2\eta} \tag{8.5}$$

$$\omega = \frac{\mu^2}{2\eta} \tag{8.6}$$

8.3 Results and discussion

8.3.1 Equilibrium geometries

For bi-metallic $Ag_2Au_N^{\lambda}$ ($N = 1$–7, $\lambda = \pm 1, 0$) clusters, the optimized ground state configurations of neutral, cationic and anionic state are shown in Figures 8.1–8.3 respectively. For Ag_2Au cluster ($N = 1$ and $\lambda = \pm 1, 0$), we have computed triangular geometrical structure with symmetry group C_{2v}. This is in line with the previous results reported by Bonacic- Koutecky et al. [40] and Tafoughalt et al. [41]. In Ag_2Au^+

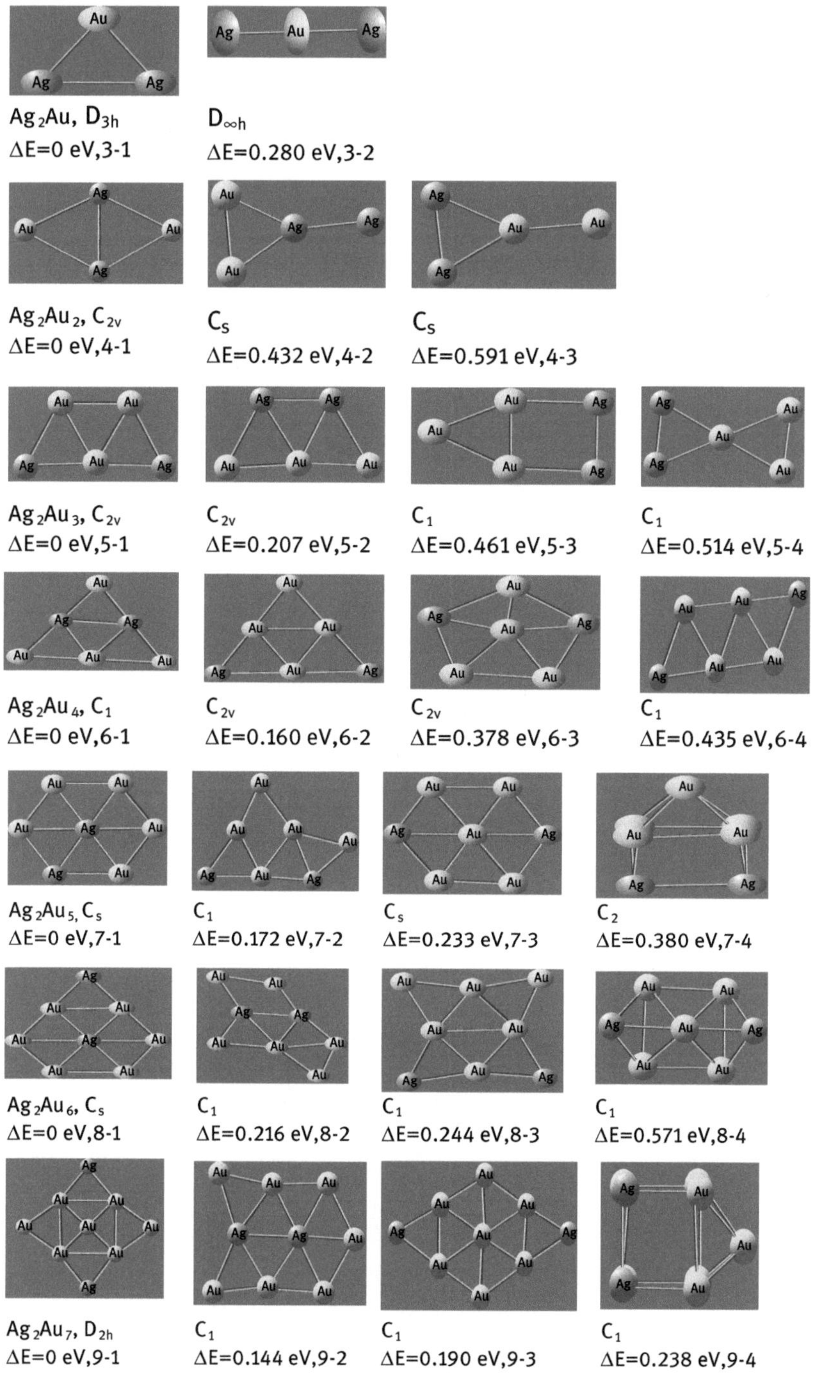

Figure 8.1: The ground state configurations of $Ag_2Au_N^{\lambda}$ ($N = 1–7$, $\lambda = 0$) nanoalloy clusters.

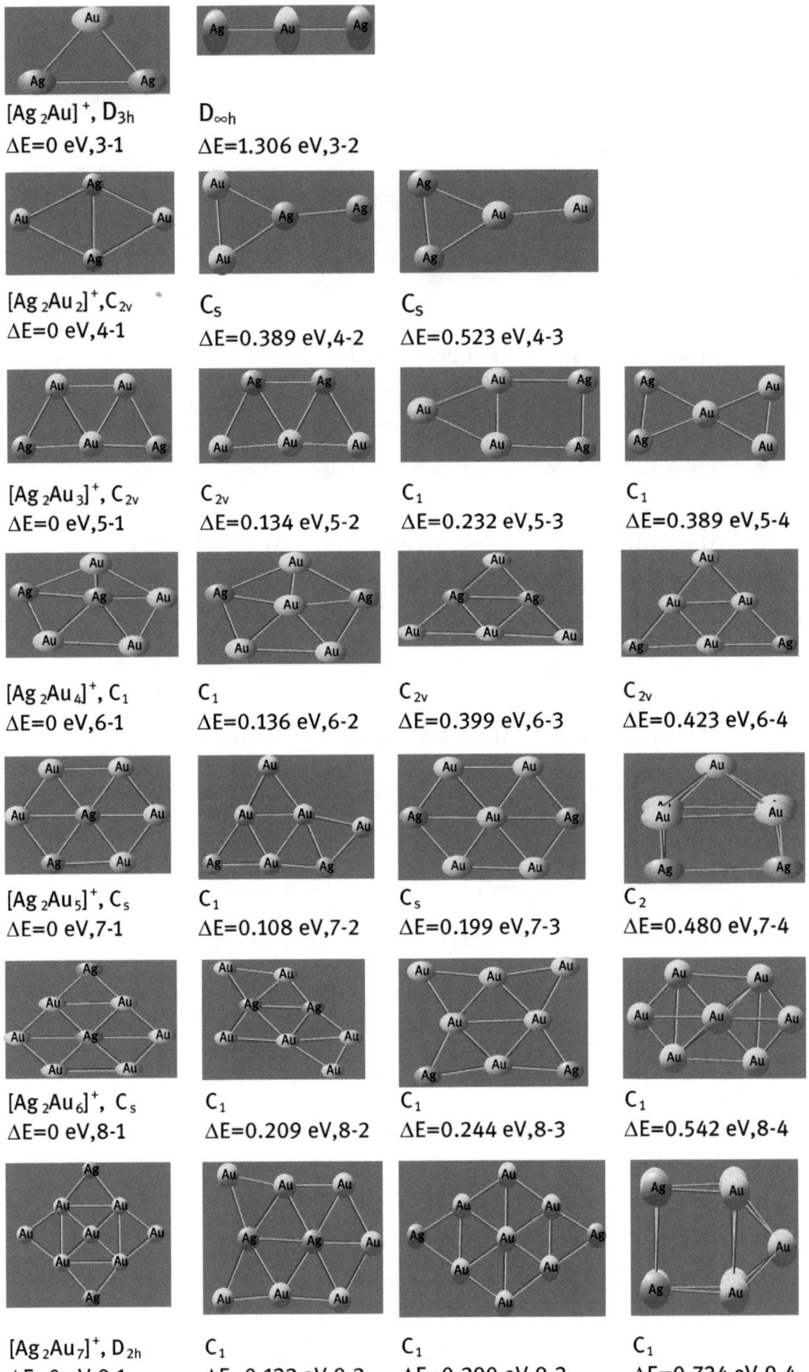

Figure 8.2: The ground state configurations of $Ag_2Au_N^{\lambda}$ ($N = 1–7$, $\lambda = +1$) nanoalloy clusters.

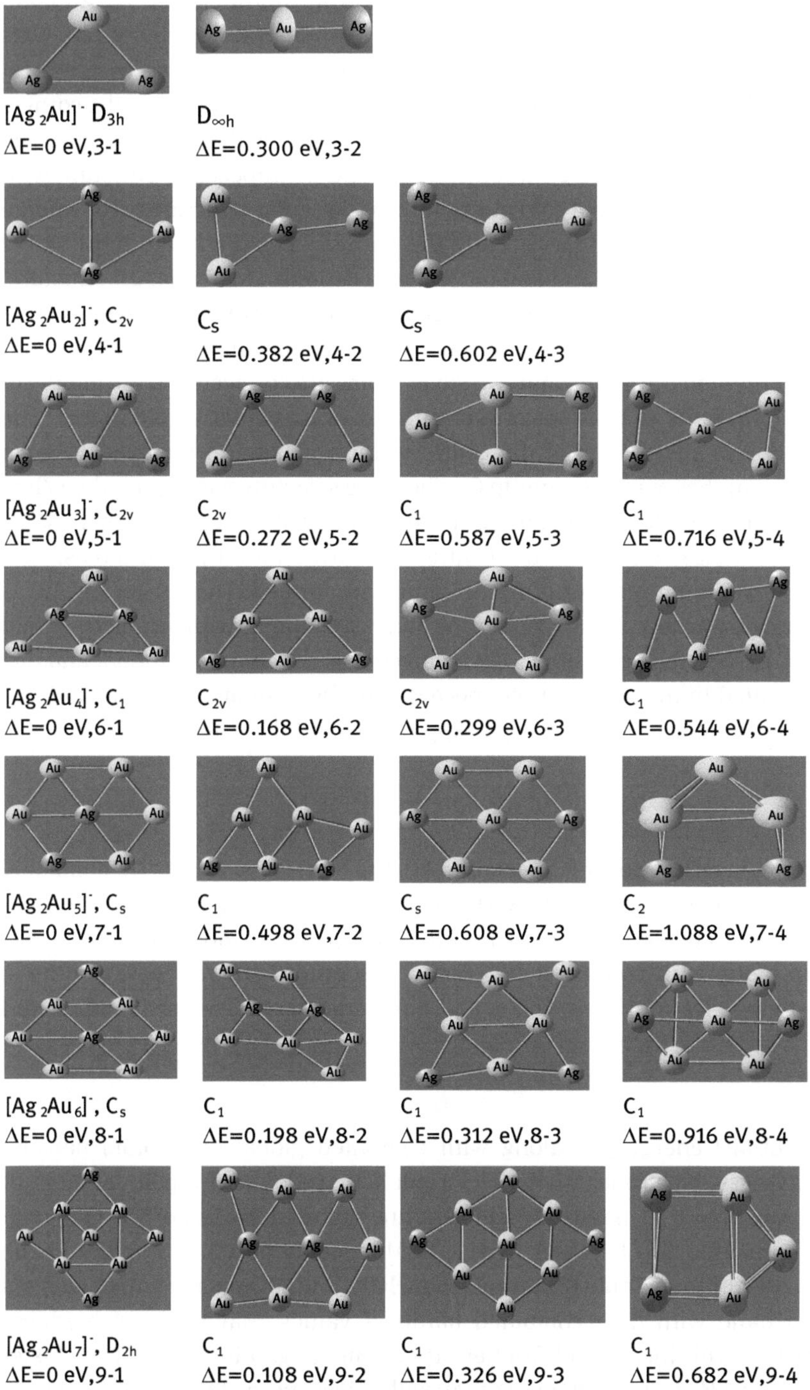

Figure 8.3: The ground state configurations of $Ag_2Au_N^{\lambda}$ ($N = 1–7$, $\lambda = -1$) nanoalloy clusters.

cluster, the Ag-Au bond length is 2.60 Å and in neutral Ag_2Au cluster, Ag-Au bond length is 2.64 Å which are in agreement with the previous values reported by Weis et al. [38]. For $N = 2$, Ag_2Au_2 cluster has rhombus structure, observed as the most stable geometry for tetramers. In this geometry, gold atoms are located at the corners of the clusters, while silver atoms connected to each other with short diagonal. The ground state configuration of Ag_2Au_3 cluster is found as trapezoidal structure with symmetry group C_{2v}. Here, three gold atoms form a triangular structure in the middle of the cluster, while silver atoms occupy the corners. For $N = 4$, the most stable structure is planar for neutral and anionic whereas cationic cluster has 2D structure. For neutral and anionic forms of Ag_2Au_4 cluster, the structures are extension of Ag_2Au_2 cluster with C_{2v} symmetry in which two gold atoms are connected on both the corners of rhombus structure and two silver atoms are connected in the middle of the cluster. It supports to form maximum number of bonds with gold atoms. The cluster Ag_2Au_5 has silver atom at the centre to form maximum number of Ag-Au bonds. The Ag_2Au_5 has symmetry group C_s. The same structure was reported by Zhao et al. [44]. The most stable structure of Ag_2Au_6 is obtained due to extension of Ag_2Au_5. Here the silver atoms, located at the centre, are surrounded by gold atoms. And consequently maximum number of Ag-Au bonds are formed. The cluster Ag_2Au_6 is under C_1 symmetry group. For $N = 7$, all nine atoms are planar with symmetry group D_{2h}. In this structure, four gold atoms are connected as a rectangular format and one gold atom, located in the middle, is connected with these atoms. Two gold and two silver atoms are connected to form edge capping from all the sides.

8.3.2 Electronic properties and DFT-based descriptors

Computational investigation of bimetallic $Ag_2Au_N^{\lambda}$ ($N = 1–7$, $\lambda = \pm 1$, 0) nanoalloy clusters has been performed invoking DFT method. The orbital energies in form of HOMO (Highest Occupied Molecular Orbital)-LUMO (Lowest Unoccupied Molecular Orbital) gap along with computed DFT-based descriptors viz. hardness, softness, electronegativity, electrophilicity index and dipole moment have been studied and analysed.

8.3.2.1 Bimetallic Ag_2Au_N ($N = 1–7$) nanoalloy clusters

The HOMO–LUMO energy gap along with computed global descriptors namely, hardness, softness, electronegativity, electrophilicity index and dipole moment of neutral Ag_2Au_N ($N = 1–7$) nanoalloy clusters are reported in Table 8.1. Our data supports the relationship between HOMO–LUMO energy gap and stability of the compounds. Result from Table 8.1 reveals that HOMO–LUMO energy gap of clusters run hand in hand with their computed hardness values. Magnitude of hardness values increases with increase of frontier orbital energy gap increases. The cluster having the highest HOMO–LUMO energy gap will be least prone towards any external perturbation. From Table 8.1, it is concluded that the Ag_2Au_4 with symmetry group C_1

Table 8.1: Computed density functional theory-based descriptors of Ag_2Au_N ($N = 1–7$) nanoalloy clusters.

Species	HOMO–LUMO Gap (eV)	Electronegativity (eV)	Hardness (eV)	Softness (eV)	Electrophilicity Index (eV)	Dipole Moment (Debye)
Ag_2Au	0.570	4.584	0.285	1.740	36.788	2.168
Ag_2Au_2	1.687	5.278	0.843	0.592	16.517	5.450
Ag_2Au_3	1.659	4.856	0.829	0.602	14.212	2.164
Ag_2Au_4	1.823	5.673	0.911	0.548	17.655	1.010
Ag_2Au_5	0.626	5.074	0.312	1.597	41.148	0.000
Ag_2Au_6	1.088	5.578	0.544	0.918	28.587	2.824
Ag_2Au_7	0.572	5.755	0.287	1.751	57.961	0.000

has maximum HOMO–LUMO energy gap (1.823 eV), whereas Ag_2Au cluster with symmetry group C_{2v} possess the lowest energy gap (0.570 eV). In absence of any quantitative data for optical properties of these clusters, we have tacitly assumed that a direct relationship between optical properties of Ag-Au clusters with their computed HOMO–LUMO energy gap. The assumption is based on the fact that optical properties of clusters are interrelated with flow of electrons within the systems, which in turn depends on the energy gap between valence and conduction band. Xiao et al. [86] have already established a relationship between HOMO–LUMO gap and energy difference of valence-conduction band. In view of that, we may conclude that optical properties of the bimetallic clusters exhibit a direct relationship hardness values and similarly, molecular softness exhibits an inverse relationship with the optical properties [87, 88]. The correlation diagram between HOMO–LUMO energy gap and computed softness is reported in Figure 8.4. High

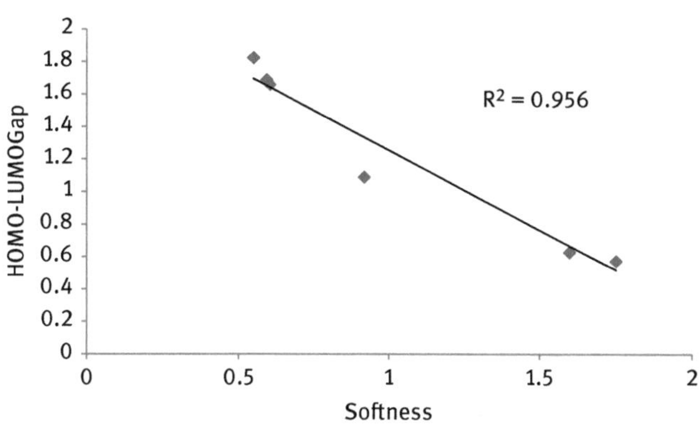

Figure 8.4: A linear correlation plot between Softness vs. HOMO–LUMO gap of Ag_2Au_N ($N = 1–7$) nanoalloy clusters.

value of regression coefficient ($R^2 = 0.956$) also establishes the relationship between HOMO–LUMO and computed softness. Electrophilicity index measures the energy lowering of a ligand due to maximal electron flow between donor and acceptor and it depends on conjoint action of ionization potential and electron affinity [89]. There is an inverse relationship HOMO–LUMP energy gap and electrophilicity index.

Dissociation energy and second difference in energy have a high impact on the relative stability of nanoalloy clusters [90–92]. These energies are highly sensitive and they exhibit an odd-even oscillation behaviour as a function of cluster size [90–92]. The HOMO–LUMO energy gap of any compound also shows an odd-even oscillation behaviour. HOMO–LUMO energy gap as a function of cluster size is depicted in Figure 8.5. Figure 8.5 displays an odd-even oscillation behaviour in terms of HOMO–LUMO energy gap. The clusters with an even number of total atoms possess larger HOMO–LUMO energy gap as compared to the neighbour clusters having odd number of total atoms. The closed-shell electronic configuration supports the stability of clusters with even number of total atoms. So, our computation runs hand in hand with reported even-odd oscillation behaviour of bimetallic nanoalloy clusters.

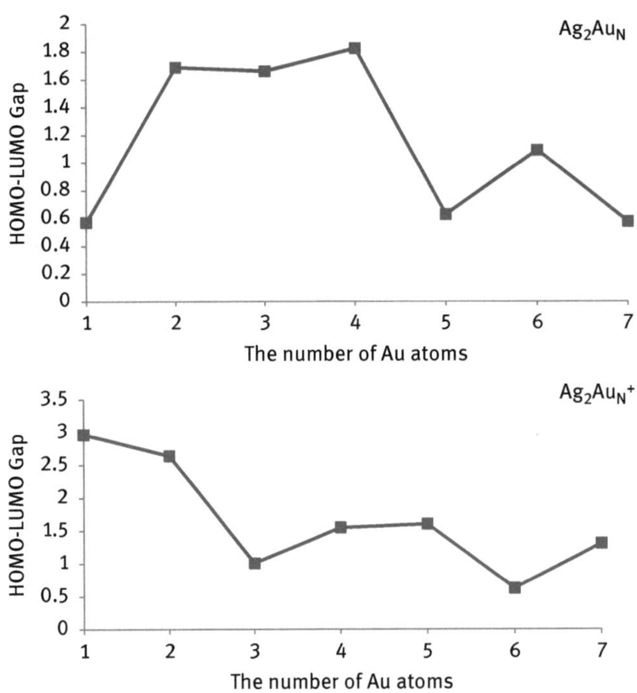

Figure 8.5: Size dependence of the HOMO–LUMO gaps for the lowest energy structures of Ag_2Au_N/ $Ag_2Au_N^+$ ($N = 1$–7) clusters.

Table 8.2: The computed bond length (in Å) for the species Ag_2, Au_2 and AgAu clusters [93–95].

Species	Computed bond length	Experimental bond length
Ag_2	2.51	2.53[93]
Au_2	2.49	2.47[94]
AgAu	2.50	2.50[95]

In order to apply our computed data in the real field, bond length of some of the species have been computed. A comparative analysis has been reported between experimental bond length [93–95] and our data in Table 8.2. A close agreement in magnitude between experimental bond length and our computation is observed from Table 8.2.

8.3.2.2 Bimetallic $Ag_2Au_N^+$(N = 1–7) nanoalloy clusters

In this section, we have discussed about the HOMO–LUMO energy gap and computed DFT-based descriptors of cationic bimetallic nanoalloy clusters. Results are reported in Table 8.3. It is observed from Table 8.3 that Ag_2Au^+ cluster, with symmetry group C_{2v}, has the maximum HOMO–LUMO energy gap whereas $Ag_2Au_6^+$ cluster, with symmetry group C_1, possesses the least energy gap. The HOMO–LUMO energy gap of $Ag_2Au_N^+$ cluster maintains a linear relationship with its hardness. However, softness and electrophilicity index exhibit an inverse relationship with the HOMO–LUMO energy gap. The linear correlation between HOMO–LUMO energy gap and computed softness of cationic $Ag_2Au_N^+$ cluster is plotted in Figure 8.6. A high regression coefficient is obtained from Figure 8.6.

The HOMO–LUMO energy gap of $Ag_2Au_N^+$ (N = 1–7) clusters is reported in Figure 8.5. Odd-even oscillation behaviour of cationic clusters is in terms of HOMO–LUMO energy gap also reported in Figure 8.5.

Table 8.3: Computed density functional theory–based descriptors of $Ag_2Au_N^+$ (N = 1–7) nanoalloy clusters.

Species	HOMO–LUMO Gap (eV)	Electronegativity (eV)	Hardness (eV)	Softness (eV)	Electrophilicity Index (eV)	Dipole Moment (Debye)
Ag_2Au^+	2.965	10.897	1.483	0.337	40.041	3.424
$Ag_2Au_2^+$	2.639	9.891	1.319	0.378	37.065	4.961
$Ag_2Au_3^+$	1.006	10.245	0.503	0.993	104.245	3.011
$Ag_2Au_4^+$	1.551	9.646	0.775	0.645	59.991	1.364
$Ag_2Au_5^+$	1.605	9.890	0.803	0.623	60.937	0.000
$Ag_2Au_6^+$	0.625	9.618	0.313	1.597	147.835	3.879
$Ag_2Au_7^+$	1.306	9.605	0.653	0.765	70.637	0.000

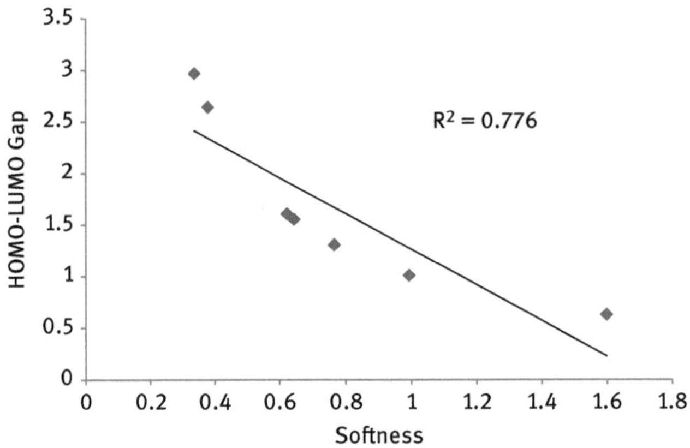

Figure 8.6: A linear correlation plot between Softness vs. HOMO–LUMO gap of $Ag_2Au_N^+$ ($N = 1–7$) nanoalloy clusters.

8.3.2.3 Bimetallic $Ag_2Au_N^-$ ($N = 1–7$) nanoalloy clusters

Anionic clusters of $Ag_2Au_N^-$ is investigated in terms of DFT-based descriptors and results are reported in Table 8.4. Table 8.4 reveals that the cluster $Ag_2Au_3^-$, with symmetry group C_{2v}, have the highest HOMO–LUMO energy gap whereas the cluster Ag_2Au^-, with symmetry group C_{2v}, possesses the lowest energy gap. The inverse relationship between HOMO–LUMO energy gap and global softness of anionic clusters is plotted in Figure 8.7. $R^2 = 0.883$ is obtained as regression coefficient.

Our computed data reveals that the magnitude of HOMO–LUMO energy gap of cationic clusters are higher than anionic counterparts with exception of $Ag_2Au_3^+$ and $Ag_2Au_6^+$ clusters.

Table 8.4: Computed density functional theory–based descriptors of $Ag_2Au_N^-$ ($N = 1–7$) nanoalloy clusters.

Species	HOMO–LUMO Gap (eV)	Electronegativity (eV)	Hardness (eV)	Softness (eV)	Electrophilicity Index (eV)	Dipole Moment (Debye)
Ag_2Au^-	0.272	0.082	0.136	3.675	0.025	0.558
$Ag_2Au_2^-$	0.762	0.054	0.381	1.312	0.004	0.000
$Ag_2Au_3^-$	1.224	0.803	0.612	0.817	0.526	0.971
$Ag_2Au_4^-$	1.034	0.816	0.517	0.967	0.644	1.346
$Ag_2Au_5^-$	0.544	1.306	0.272	1.837	3.134	0.000
$Ag_2Au_6^-$	0.979	1.605	0.489	1.021	2.631	2.018
$Ag_2Au_7^-$	0.275	0.077	0.137	3.638	0.021	0.000

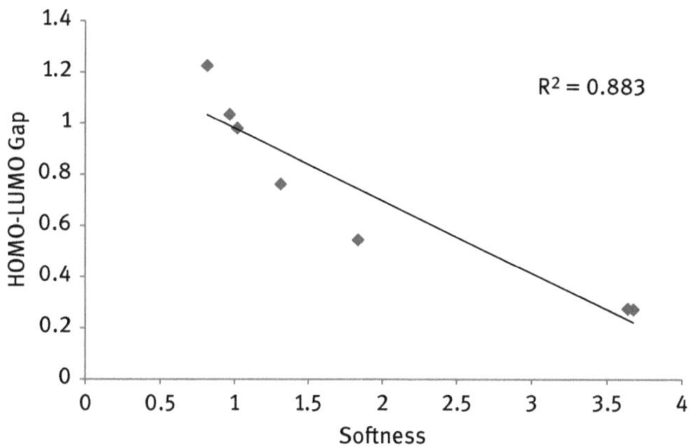

Figure 8.7: A linear correlation plot between Softness vs. HOMO–LUMO gap of $Ag_2Au_N^-$ ($N = 1–7$) nanoalloy clusters.

8.4 Conclusion

Due to diverse applications in the real field, the study of bimetallic nanoalloy clusters has gained a significant importance. In this report, we have investigated the structure and physico-chemical properties of $Ag_2Au_N^{\lambda}$ ($N = 1–7$, $\lambda = \pm1,0$) clusters in terms of DFT study. DFT-based global descriptors namely, HOMO–LUMO energy gap, hardness, softness, electronegativity, electrophilicity index and dipole moment of cationic, anionic and neutral Ag_2Au_N clusters are computed and reported. The result reveals a direct relationship between HOMO–LUMO energy gap with global hardness and inverse relationship with softness. Considering optical property of clusters as a function of energy gap, we have tried to explain optical properties/HOMO–LUMO energy gap in terms of computed descriptors. Our predicted trend is also in similar line with experimental facts. Odd-even oscillation behaviour of neutral and charged clusters are also established invoking our computed data. The neutral clusters with even number of total atoms possess larger energy gap as compared to their neighbour clusters with odd number of total atoms. It is concluded that Ag_2Au^+ cluster is the most stable cluster among all molecular system. Our computed bond length for the species of Ag_2, Au_2 and Ag-Au are in well agreement with the experimental values. It strengthens our computational analysis.

References

[1] Khosousi AZ, Dhirani AA. Charge transport in nanoparticle assemblies. Chem Rev 2008;108:4072.
[2] Daniel MC, Astruc D. Gold nanoparticles: assembly, supramolecular chemistry, quantum-size-related properties, and applications toward biology, catalysis and nanotechnology. Chem Rev 2004;104:293.

[3] Ghosh SK, Pal T. Photophysical aspects of molecular probes near nanostructured gold surfaces. Chem Rev 2007;107:4797.

[4] Chaudhuri RG, Paria S. Core/shell nanoparticles: classes, properties, synthesis mechanisms, characterization, and applications. Chem Rev 2012;112:2373.

[5] Alivisatos AP. Semiconductor clusters, nanocrystals and quantum dots. Science. 1996;271:933.

[6] Wang HQ, Kuang XY, Li HF. Density functional study of structural and electronic properties of bimetallic copper-gold clusters: comparison with pure and doped gold clusters. Phys Chem Chem Phys 2010;12:5156.

[7] Roucoux A, Schulz J, Patin H. Reduced transition metal colloids: a novel family of reusable catalysts? Chem Rev 2002;102:3757.

[8] Rao CN, Cheetham AK. Science and technology of nanomaterials: current status and future prospects. J Mater Chem 2001;11:2887.

[9] Munoz-Flores BM, Kharisov BI, Jimenez-Perez VM, Martinez PE, Lopez ST. Recent advances in the synthesis and main applications of metallic nanoalloys. Ind Eng Chem Res 2011;50:7705.

[10] Murray RW. Nanoelectrochemistry: metal nanoparticles, nanoelectrodes and nanopores. Chem Rev 2008;108:2688.

[11] Yin F, Wang ZW, Palmer RE. Formation of bimetallic nanoalloys by Au coating of size selected Cu clusters. J Nanopart Res 2012;14:1124.

[12] Teng X, Wang Q, Liu P, Han W, Frenkel AI, Wen MN, et al. Formation of Pd/Au nanostructures from Pd nanowires via galvanic replacement reaction. J Am Chem Soc 2008;130:1093.

[13] Ferrando R, Jellinek J, Johnston RL. Nanoalloys: from theory to applications of alloy clusters and nanoparticles. Chem Rev 2008;108:845.

[14] Henglein A. Physicochemical properties of small metal particles in solution: “microelectrode” reactions, chemisorption, composite metal particles, and the atom-to-metal transition. J Phys Chem 1993;97:5457.

[15] Davis SC, Klabunde KJ. Unsupported small metal particles: preparation, reactivity and characterization. Chem Rev 1982;82:153.

[16] Lewis LN. Chemical catalysis by colloids and clusters. Chem Rev 1993;93:2693.

[17] Schmid G. Large clusters and colloids. Metals in the embryonic state. Chem Rev 1992;92:1709.

[18] Schon G, Simon U. A fascinating new field in colloid science: small ligand-stabilized metal clusters and possible applications in microelectronics. Colloid Polym Sci 1995;273:101.

[19] Oderji HY, Ding H. Determination of melting mechanism of $Pd_{24}Pt_{14}$ nanoalloy by multiple histogram method via molecular dynamics simulations. Chem Phys 2011;388:23.

[20] Huang X, Sayed IE, Qian W, Sayed MA. Cancer cell imaging and photothermal therapy in the near-infrared region by using gold nanorods. J Am Chem Soc 2006;128:2115.

[21] Hainfeld J, Slatkin D, Smilowitz HM. The use of gold nanoparticles to enhance radiotherapy in mice. Phys Med Biol 2004;49:309.

[22] Wu D, Zhang XD, Liu PX, Zhang LA, Fan FY, Guo ML. Gold nanostructure: fabrication, surface modification, targeting imaging, and enhanced radiotherapy. Curr Nanosci 2011;7:110.

[23] Zhang X, Guo M, Wu D, Liu P, Sun Y, Zhnag L, et al. First-principles investigation of Ag-doped gold nanoclusters. Int J Mol Sci 2011;12:2972.

[24] Chen J, Wang D, Xi J, Au L, Siekkinen A, Warsen A, et al. Immuno gold nanocages with tailored optical properties for targeted photothermal destruction of cancer cells. Nano Lett 2007;7:1318.

[25] Robinson JM, Takizawa T, Vandre DD. Applications of gold cluster compounds in immunocytochemistry and correlative microscopy: comparison with colloidal gold. J Microsc 2000;199:163.

[26] Whetten RL, Khoury JT, Alvarez MM, Murthy S, Vezmar I, Wang ZL, et al. Nanocrystal gold molecules. Adv Mater 1996;5:428.

[27] Andres RP, Bein T, Dorogi M, Feng S, Henderson JI, Kubiak CP, et al. “Coulomb Staircase” at room temperature in a self-assembled molecular nanostructure. Science 1996;272:1323.

[28] Mirkin CA, Letsinger RL, Mucic RC, Storhoff JJ. A DNA-based method for rationally assembling nanoparticles into macroscopic materials. Nature 1996;382:607.

[29] Alivisatos AP, Johnsson KP, Peng XG, Wilson TE, Loweth CJ, Bruchez MP, et al. Organization of ‘nanocrystal molecules’ using DNA. Nature 1996;382:609.

[30] Andres RP, Bielefeld JD, Henderson JI, Janes DB, Kolagunta VR, Kubiak CP, et al. Self-assembly of a two-dimensional superlattice of molecularly linked metal clusters. Science 1996;273:1690.

[31] Garzon IL, Michaelian K, Beltran MR, Amarillas AP, Ordejon P, Artacho E, et al. Lowest energy structures of gold nanoclusters. Phys Rev Lett 1998;81:1600.

[32] Garzon IL, Rovira C, Michaelian K, Beltran MR, Ordejon P, Junquera J, et al. Do thiols merely passivate gold nanoclusters? Phys Rev Lett 2000;85:5250.

[33] Michaelian K, Rendon N, Garzon IL. Structure and energetics of Ni, Ag, and Au nanoclusters. Phys Rev B 1999;60:2000.

[34] Schaaff TG, Whetten RL. Giant gold-glutathione cluster compounds: intense optical activity in metal-based transitions. J Phys Chem B 2000;104:2630.

[35] Priyanka, Dharamvir K. The structural and electronic properties of tubular gold clusters with a spinal support. Phys Chem Chem Phys 2013;15:12340.

[36] Negishi Y, Nakamura Y, Nakajima A, Kaya K. Photoelectron spectroscopy of gold-silver binary cluster anions (Au_nAg_{m-}; 2≤n+m≤4). J Chem Phys 2001;115:3657.

[37] Heiles S, Logsdail AJ, Schfer R, Johnston RL. Dopant-induced 2D-3D transition in small Au-containing clusters: DFT-global optimization of 8-atom Au-Ag nanoalloys. Nanoscale 2012;4:1109.

[38] Weis P, Welz O, Vollmer E, Kappes MM. Structures of mixed gold-silver cluster cations (Ag_mAu_{n+}, m+n<6): ion mobility measurements and density-functional calculations. J Chem Phys 2004;120:677.

[39] Lee HM, Ge M, Sahu BR, Tarakeshwar P, Kim KS. Geometrical and electronic structures of gold, silver and gold-silver binary clusters: origins of ductility of gold and gold-silver alloy formation. J Phys Chem B 2003;107:9994.

[40] Kouteck VB, Burda J, Mitrić R, Ge M. Density functional study of structural and electronic proeprties of bimetallic silver-gold clusters: comparison with pure gold and silver clusters. J Chem Phys 2002;117:3120.

[41] Tafoughalt MA, Samah M. Density functional investigation of structural and electronic properties of small bimetallic silver-gold clusters. Phys B 2012;407:2014.

[42] Shayeghi A, Heard CJ, Johnston RL, Schäfer R. Optical and electronic properties of mixed Ag-Au tetramer cations. J Chem Phys 2014;140:054312.

[43] Zhao YR, Kuang XY, Zheng BB, Li YF, Wang SJ. Equilibrium geometries, stabilities, and electronic properties of the bimetallic M2 doped Aun (M=Ag, Cu; n=1-10) clusters: comparison with pure gold clusters. J Phys Chem A 2011;115:569.

[44] Zhao GF, Zeng Z. Geometrical and electronic structures of AumAgn (2≤m+n≤8). J Chem Phys 2006;125:014303.

[45] Miranda FM, Menziani MC, Pedone A. Influence of silver doping on the photoluminescence of protected AgnAu25-n nanoclusters: A time-dependent density functional theory investigation. J Phys Chem C 2015;119:10766.

[46] Tafoughalt MA, Samah M. Structural properties and relative stability of silver-doped gold clusters AgAun-1 (n=3-13): density functional calculations. Comput Theor Chem 2014;1033:23.

[47] Shayeghi A, Schfer R, Rayner DM, Johnston RL, Fielicke A. Charge-induced dipole vs. relativistically enhanced covalent interactions in Ar-tagged Au-Ag tetramers and pentamers. J Chem Phys 2015;143:024310.
[48] Chi Y, Zhao L, Lu X, An C, Guo W, Wu CM. Effect of alloying on the stabilities and catalytic properties of Ag-Au bimetallic subnanoclusters: A theoretical investigation. J Mater Sci 2016;51:5046.
[49] Cramer CJ, Truhlar DG. Density functional theory for transition metals and transition metal chemistry. Phys Chem Chem Phys 2009;11:10757.
[50] Hafner J, Wolverton C, Ceder G. Toward computational materials design: the impact of density functional theory on materials research. MRS Bull 2006;31:659.
[51] Parr RG, Yang W. Density-functional theory of the electronic structure of molecules. Annu Rev Phy Chem 1995;46:701.
[52] Kohn W, Becke AD, Parr RG. Density functional theory of electronic structure. J Phys Chem 1996;100:12974.
[53] Liu S, Parr RG. J Chem Phys 1997;106:5578.
[54] Ziegler T. Chem Rev 1991;91:651.
[55] Parr RG, Yang W. Density functional theory of atoms and molecules. Oxford: Oxford University Press, 1989.
[56] Chermette H. Chemical reactivity indexes in density functional theory. J Comput Chem 1999;20:129.
[57] Geerlings P, Proft FD, Langenaeker W. Conceptual density functional theory. Chem Rev 2003;103:1793. Washington, DC.
[58] Geerlings P, Proft FD. Chemical reactivity as described by quantum chemical methods. Int J Mol Sci 2002;3:276.
[59] Ranjan P, Dhail S, Venigalla S, Kumar A, Ledwani L, Chakraborty T. A theoretical analysis of bimetallic (Cu-Ag)n=1-7 nano alloy clusters invoking DFT based descriptors. Mater Sci Pol 2015;33:719.
[60] Ranjan P, Venigalla S, Kumar A, Chakraborty T. Theoretical study of bi-metallic AgmAun; (m+n=2-8) nano alloy clusters in terms of DFT based descriptors. New Front Chem 2014;23:111.
[61] Venigalla S, Dhail S, Ranjan P, Jain S, Chakraborty T. Computational study about cytotoxicity of metal oxide nanoparticles invoing nano-QSAR technique. New Front Chem 2014;23:123.
[62] Ranjan P, Kumar A, Chakraborty T. Computational study of AuSin (n=1-9) nanoalloy clusters invoking DFT based descriptors. AIP Conf Proc 2016;1724:020072.
[63] Ranjan P, Kumar A, Chakraborty T. Theoretical analysis: electronic and optical proeprties of gold-silicon nanoalloy clusters. Mat Today Proc 2016;3:1563.
[64] Ranjan P, Kumar A, Chakraborty T, Mishra GC. Computational study of nanomaterials invoking DFT based descriptors. Environmental sustainability: concepts, principles, evidences and innovations. New Delhi: Excellent Publishing House, 2014:239–242.
[65] Ranjan P, Venigalla S, Kumar A, Chakraborty T, Chakraborty T, Ledwani L. A theoretical analysis of bi-metallic Ag-Aun;(n=1-7) nano alloy clusters invoking DFT based descriptors. Recent methodology in chemical sciences: experimental and theoretical approaches. USA: Apple Academic Press and CRC Press, 2015:337–346.
[66] Ranjan P, Kumar A, Chakraborty T. Computational investigation of Ge doped Au nanoalloy clusters: A DFT study. IOP Conf Series Mater Sci Eng 2016;149:012172.
[67] Dhail S, Ranjan P, Chakraborty T, Ramasami P, Bhowon MG, Laulloo SJ, et al. Correlation of the experimental and theoretical study of some novel 2-phenazinamine derivatives in terms of DFT-based descriptors. Crystallizing ideas- the role of chemistry. Switzerland: Springer, 2016:97–112.

[68] Ranjan P, Kumar A, Chakraborty T, Chakraborty T, Ranjan P, Pandey A. Theoretical analysis: electronic and optical properties of small Cu-Ag nano alloy clusters. Computational chemistry methodology in structural biology and material sciences. USA: Apple Academic Press and CRC Press, In Press. ISBN- 9781315207544.
[69] Ranjan P, Chakraborty T, Kumar A, Haghi AK, Pogilani L, Castro EA, et al. A theoretical study of bimetallic CuAun (N=1-7) nanoalloy clusters invoking conceptual DFT based descriptors. Applied chemistry and chemical engineering Vol. 4. USA: Apple Academic Press and CRC Press, In Press. ISBN- 9781315207636.
[70] Frisch MJ, Trucks GW, Schlegel HB, Scuseria GE, Robb MA, Cheeseman JR, et al. Gaussian 03, Revision C.02. Wallingford CT: Gaussian, Inc., 2004.
[71] Jones RO, Gunnarsson O. The density functional formalism, its applications and prospects. Rev Mod Phys 1989;61:689.
[72] Zupan A, Blaha P, Schwarz K, Perdew JP. Pressure-induced phase transitions in solid Si, Sio2, and Fe: performance of local-spin-density and generalized-gradient-approximation density functionals. Phys Rev B 1998;58:11266.
[73] Theilhaber J. Quantum-molecular-dynamics simulations of liquid metals and highly degenerate plasmas. Phys Fluids B 1992;4:2044.
[74] Stadler R, Gillan MJ. First-principles molecular dynamics studies of liquid tellurium. J Phys Condens Matter 2000;12:6053.
[75] Argaman N, Makov G. Density functional theory-a introduction. Am J Phys Condens Matter 2000;68:69.
[76] Chen FY, Johnston RL. Structure and spectral characteristics of the nanoalloy Ag3Au10. Appl Phys Lett 2007;90:153123.
[77] Chen FY, Johnston RL. Charge transfer driven surface segregation of gold atoms in 13-atom Au-Ag nanoalloys and its relevance to their structural, optical and electronic properties. Acta Mater 2008;56:2374.
[78] Jug K, Zimmermann B, Calaminici P, Köster AM. Structure and stability of small copper clusters. J Chem Phys 2002;116:4497.
[79] Hay PJ, Wadt WR. Ab initio effective core potentials for molecular calculations. Potentials for the transition metal atoms Sc to Hg. J Chem Phys 1985;82:299.
[80] Assadollahzadeh B, Schwerdtfeger P. A systematic search for minimum structures of small gold clusters Aun (n=2-20) and their electronic properties. J Chem Phys 2009;131:064306.
[81] Chiodo S, Russo N, Sicilia E. LANL2DZ basis sets recontracted in the framework of density functional theory. J Chem Phys 2006;125:104107.
[82] Jiang ZY, Lee KH, Li ST, Chu SY. Structures and charge distributions of cationic and neutral Cun-1Ag clusters (n=2-8). Phys Rev B 2006;73:235423.
[83] Mielich B, Savin A, Stoll H, Preuss H. Results obtained with the correlation energy density functionals of Becke and Lee, Yang and Parr. Chem Phys Lett 1989;157:200.
[84] Petrie S. Pitfalls for the frozen-core approximation: Gaussian-2 calculations on the sodium cation affinities of diatomic fluorides. J Phys Chem A 1998;102:6138.
[85] Purwanto W, Zhang S, Krakauer H. Frozen-orbital and downfolding calculations with auxiliary-field quantum Monte Carlo. J Chem Theory Comput 2013;9:4825.
[86] Xiao H, Kheli JT, Goddard WA. Accurate band gaps for semiconductors from density functional theory. J Phys Chem Lett 2011;2:212.
[87] Parr RG. Absolute electronegativity and hardness correlated with molecular orbital theory. Proc Natl Acad Sci 1986;83:8440.
[88] Kaya Y, Yilmaz VT, Buyukgungor O. Synthesis, spectroscopic, structural and quantum chemical studies of a new imine oxime and its palladium (II) complex: hydrolysis mechanism. Molecules 2016;21:52.

[89] Parr RG, Szentpaly LV, Liu S. Electrophilicity index. J Am Chem Soc 1999;121:1929.
[90] Ding LP, Kuang XY, Shao P, Zhao YR, Li YF. A comparative study on geometries, stabilities and electronic properties between bimetallic AgnX (X=Au, Cu; n=1-8) and pure silver clusters. Chin Phys B 2012;21:043601.
[91] Hakkinen H, Landman U. Gold clusters (AuN, 2<N<10). Phys Rev B 2000;62:2287.
[92] Li XB, Wang HY, Yang XD, Zhu ZH, Tang YJ. Size dependence of the structures and energetic and electronic properties of gold clusters. J Chem Phys 2007;126:084505.
[93] Beutel V, Kramer HG, Bhale GL, Kuhn M, Weyers K, Demtroder W. High-resolution isotope selective laser spectroscopy of Ag2 molecules. J Chem Phys 1993;98:2699.
[94] Huber KP, Herzberg G. Constraints of diatomic molecules. New York: Van Nostrand Reinhold Company, 1979.
[95] Fabbi JC, Langenberg JD, Costello QD, Morse MD, Karlsson L. Dispersed fluorescence spectroscopy of AlNi, NiAu and PtCu. J Chem Phys 2001;115:7543.

N Chooramun, P. J. Lawrence and E. R. Galea

9 Evacuation simulation using Hybrid Space Discretisation and Application to Large Underground Rail Tunnel Station

Abstract: In all evacuation simulation tools, the space through which agents navigate and interact is represented by one the following methods, namely Coarse regions, Fine nodes and Continuous regions. Each of the spatial representation methods has its benefits and limitations. For instance, the Coarse approach allows simulations to be processed very rapidly, but is unable to represent the interactions of the agents from an individual perspective; the Continuous approach provides a detailed representation of agent movement and interaction but suffers from relatively poor computational performance. The Fine nodal approach presents a compromise between the Continuous and Coarse approaches such that it allows agent interaction to be modelled while providing good computational performance. Our approach for representing space in an evacuation simulation tool differs such that it allows evacuation simulations to be run using a combination of Coarse regions, Fine nodes and Continuous regions. This approach, which we call Hybrid Spatial Discretisation (HSD), is implemented within the buildingEXODUS evacuation simulation software. The HSD incorporates the benefits of each of the spatial representation methods whilst providing an optimal environment for representing agent movement and interaction. In this work, we demonstrate the effectiveness of the HSD through its application to a moderately large case comprising of an underground rail tunnel station with a population of 2,000 agents.

Keywords: evacuation modelling using hybrid space, Hybrid Spatial Discretisation, buildingEXODUS-Hybrid, hybrid evacuation simulation

9.1 Introduction

An important feature in all evacuation simulation tools is the modelling of the space in which the agents navigate and interact with their surroundings as well as with other agents. The two methods typically used for spatial representation are discrete and continuous spaces [1]. The discretised spatial method involves the segmentation of the available space into sub-regions commonly known as nodes, and each node can be connected to its neighbours via arcs or links. This method allows the representation of paths or routes which can be taken by occupants to move from one location to another and can be applied to solve complex geometrical problems. The

https://doi.org/10.1515/9783110467215-009

discrete spatial method can be further divided into two categories namely Coarse region and Fine node. Coarse regions are much bigger in size such that each Coarse region can represent an entire room or segment of the space, whereas in the Fine node method, multiple nodes are used to represent a single room. In the Continuous region method, space is represented as a polygon of any geometrical shape and there are no nodes as such.

Each of the spatial representation approaches has its relative benefits and limitations. In the Coarse region approach, the occupants move from segment to segment such that they cannot be individually located. The Coarse region approach simplifies the building layout and makes it easier to manage large geometries and since the location and interaction of the occupants are not modelled, it can handle large population sizes at relatively low computational cost. In the Continuous region approach, the occupants can be placed anywhere in the geometry and have unique locations. The agent movement towards an exit or any other location of interest such as assembly point requires steering towards to a target, however, with increasing agents and obstacles, the steering process becomes increasingly difficult, especially continuously searching for nearby agents and frequent updates are required. Therefore, simulations run using the Continuous region approaches tend to be computationally expensive and thus not suitable for representing large population sizes in large geometries. In the Fine node approach, each occupant occupies a node and therefore can be uniquely identified. As such, the Fine node approach allows accurate representation of agent movement and behaviours. The Fine model represents the agent locations with less accuracy than the Continuous model. Also, since the Fine model represents people as individuals, it can still handle complex behaviours better than the Coarse model, but not to the same positional accuracy as the Continuous model [2].

The novelty of the Hybrid Spatial Discretisation (HSD)[3], presented here, is that it encompasses the benefits of Coarse regions, Continuous regions and Fine nodes within a single integrated software tool. The HSD approach is implemented in the buildingEXODUS software [4]. In this work, the HSD approach is applied to a large underground rail tunnel station with a population of 2,000 agents. The application of the HSD to tunnel evacuation is to show the usefulness and effectiveness of the technique. The HSD approach allows the mixing of spatial representation techniques that are situated on different levels on the scale of granularity. As such, the HSD can generate results on various levels of refinement, as required by the user or depending on the evacuation phenomena that is required to be modelled. For instance, Coarse regions are best for areas of uniform and homogeneous behaviour, Fine nodes for areas which may have more complex behaviour (i. e. junctions, stairs, escalators, queues) and Continuous regions where predictive flows through narrow openings are required or the best qualitative movement is required.

Using the HSD approach, enclosures can be represented using different configurations namely All-Coarse (using only Coarse regions), All-Fine (using only Fine nodes), All-Continuous (using only Continuous regions) and Hybrid (using different combinations of Fine nodes, Coarse regions and Continuous regions). Figure 9.1 illustrates the representation of multiple compartment geometry with two external exits using different spatial configurations [3].

Moreover, the work presented here facilitates the application of evacuation modelling by allowing simulations results to be obtained faster than real time rate while providing an integrated environment for representing agent movement and behaviours in constrained environments.

9.2 Software architecture for Hybrid Spatial Discretisation

The buildingEXODUS software was selected as a development and deployment platform for the HSD approach. buildingEXODUS is an evacuation simulation tool which utilises Fine nodes for representing physical space within enclosures. The movement and behaviour of the occupants is defined by a series of heuristics or rules which are grouped into 6 interacting sub-core models [5] namely OCCUPANT, MOVEMENT, BEHAVIOUR, TOXICITY, HAZARD and ENCLOSURE, as shown in Figure 9.2. The Occupant sub-model describes an individual as a collection of attributes and influences the behaviour of the occupants during the simulation. The Movement sub-model controls the movement of the occupants from their current position to the most favourable neighbouring location until they reach their final location. The Behaviour sub-model determines the response of the occupants based on their personal attributes. The Hazard sub-model controls the atmospheric and physical environment and distributes pre-determined fire hazards such as heat, smoke and toxic products. The Toxicity sub-model determines the effects on individuals exposed to toxic products (distributed by the Hazard sub-model). These effects are passed on to the Behaviour sub-model which influences the movement of the occupants. Additional details on the interacting sub-models can be found in the EXODUS manual [5].

In order to implement the HSD approach, the buildingEXODUS model was updated with a modular architecture to allow new functionalities to be incorporated into the core software. The Coarse region and Continuous region were independently developed as plug-in modules, tested and integrated into the new model, which is identified as the buildingEXODUS-Hybrid or bEX-H [3]. The Coarse region and Continuous region components are described briefly below.

9.2.1 Continuous region component

The Continuous approach for evacuation simulation makes it possible to consider a larger number of agent attributes and consequently, a more diverse range of agent behaviours to be modelled. The Continuous region component in bEX-H is based on a multi-agent system whereby each agent is modelled as an autonomous agent that is

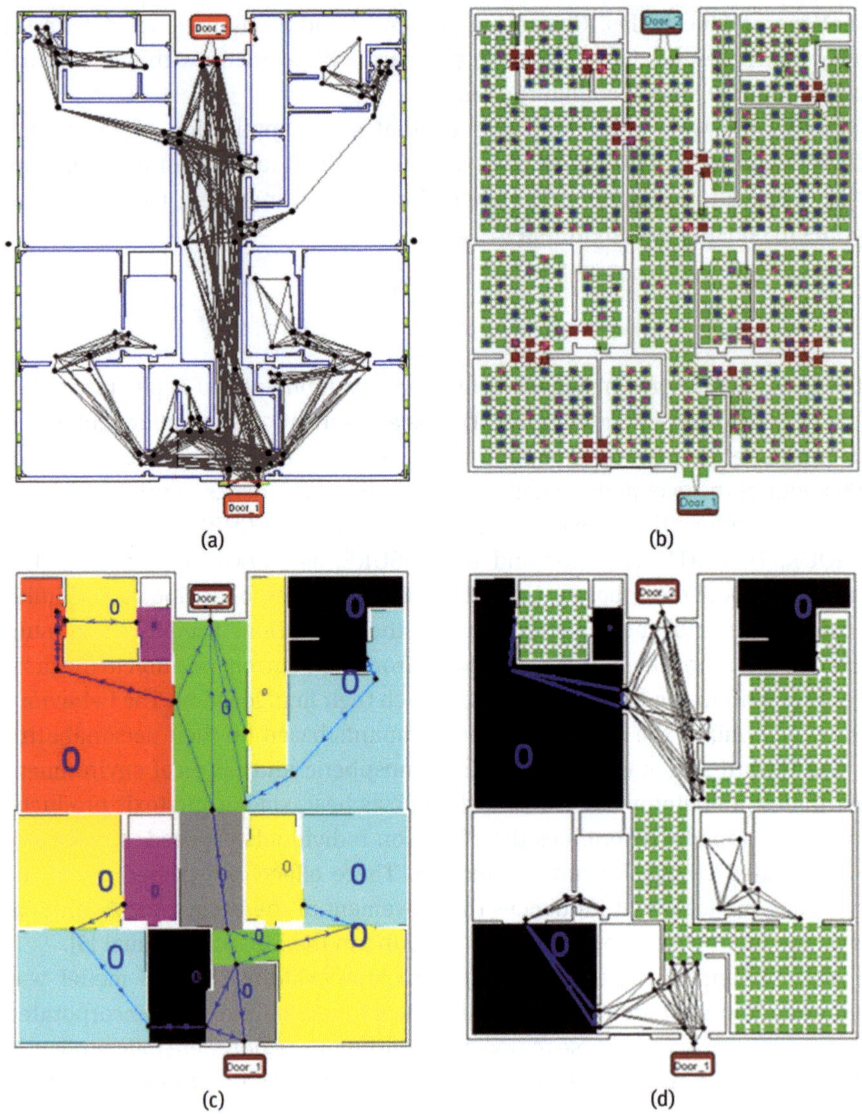

Figure 9.1: Representing a multiple compartment geometry using different configurations [3]. (a) All-Continuous – entire geometry represented using continuous space. Also shown here is the Navigational graph, an underlying data structure to facilitate navigation of agents in continuous space [3] (b) All-Fine – entire geometry represented using a network of fine nodes (c) All-Coarse – entire geometry using coarse regions. (d) Hybrid – entire geometry represented using a combination of Coarse regions, Fine nodes and Continuous regions.

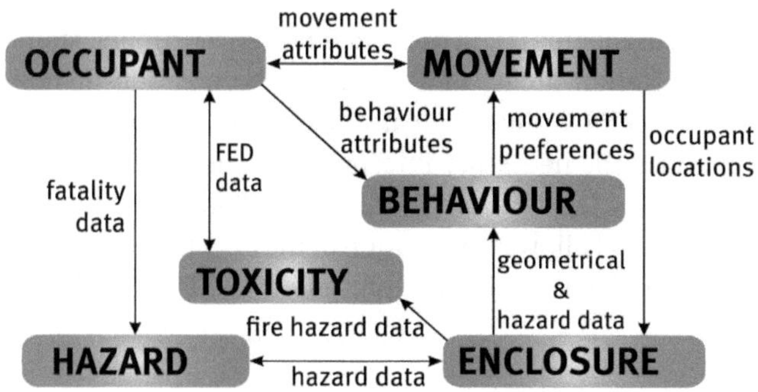

Figure 9.2: Interacting models in buildingEXODUS [5].

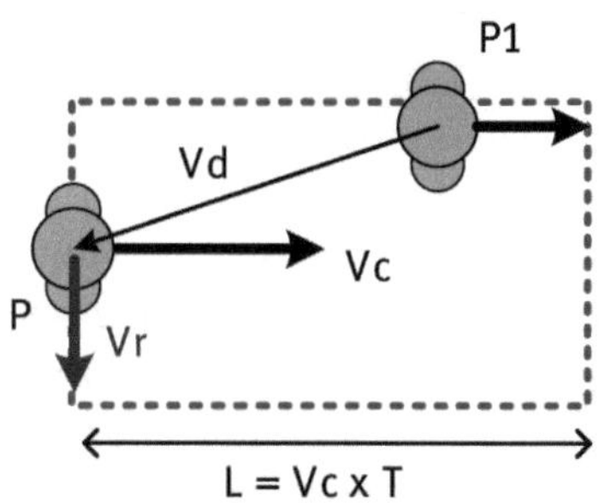

Figure 9.3: Using perception box (enclosed by dotted lines) to detect neighbouring agents.

able to adapt its behaviour and movement depending on its surroundings [3]. The Continuous region also incorporates a Navigational Graph, comprising of waypoints and arcs, to guide the agents towards their target locations [3]. This graph also facilitates the connectivity of the Continuous regions and the other spatial types namely Fine nodes and Coarse regions, when using the HSD approach.

9.2.2 Coarse region component

When using the Coarse regions, the available physical space can be segmented into partitions whereby each partition can represent a room, corridor or any other section of the geometry. All the different partitions are linked together using arcs. Each Coarse region has a maximum agent capacity while the arcs have a maximum flow capacity (maximum number of agents that can pass through the arc per unit time). The agents within the Coarse regions navigate from one segment to another while the location and agent movement within the segment itself is not represented. The flow rates through the Coarse regions are calculated based on a Flow to Density equations

[6]. The Coarse region component also includes different types of coarse segments, each with different speed-flow relationships [7], in order to represent corridors, stairs, platforms, escalators and other station areas. bEX-H allows the creation of non-convex Coarse regions. Such regions automatically generate an underlying Navigational Graph which provides the necessary information to model the flow of occupants within the Coarse regions [3]. This graph in conjunction with the algorithmic procedures in the Coarse model allows the location of the agents to be tracked when they have to transit to adjacent regions represented by Fine nodes or Continuous regions.

9.2.3 Choice of discretisation strategies for using HSD

In order to use the HSD in an optimal way, it is recommended to choose a suitable discretisation strategy, that is, a strategy for positioning the different spatial types across different sections of the geometry. For instance, the Coarse region approach is most suitable for representing sections of the geometry where agents are expected to show homogeneous behaviours (for instance, behaviour of agents moving in the same direction in a rail tunnel); Fine nodes are appropriate for representing individual agent movement and complex behaviours such as overtaking, queuing and obstacle avoidance (for example, behaviour of agents on a railway platform); Continuous regions are suitable for predicting flow rates across narrow openings and for demonstrating variation of flow rates due to small changes in geometrical dimensions (for instance, for modelling behaviour of agents near internal and external exits). More detailed guidelines and recommendations for the choice of discretisation strategies have been covered in our previous works [2, 8].

9.3 Large tunnel station complex case

This section describes the application of the HSD approach to a large tunnel station complex depicted in Figure 9.4. This geometry was chosen to demonstrate the capability of the hybrid approach to handle large complex structures such as tunnel stations. This scenario depicts passengers who have alighted from a train inside a

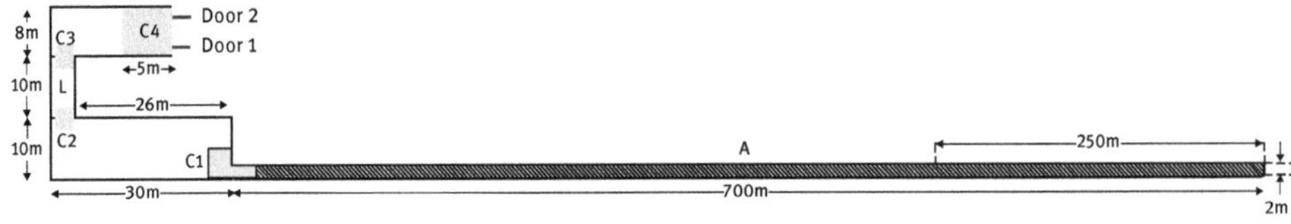

Figure 9.4: Large Tunnel Station Complex modelled using all spatial representation types; Coarse region approach (A), Continuous region (C1,C2,C3,C4), Fine node approach (Remainder of free space).

tunnel in the event of an emergency. The passengers then start to evacuate through the tunnel till they reach the platform, following which, they go through the internal exits until they reach the end of the platform. The entire geometry has a total free floor space area of 1,900 m^2 and includes a 700 m long and 2 m wide tunnel connected to a platform. As illustrated in Figure 9.4, the segment L represents a corridor of length 10 m and width 4 m and comprises of a centrally placed 2 m wide internal exit on either side of the corridor. The end of the platform is connected to two external exits, each of width 2 m. The rail tunnel station presented is modelled using a combination of all the three spatial representation approaches namely Coarse regions, Fine nodes and Continuous regions.

The spatial discretisation strategy for this geometry is chosen such that region in the far field is represented using Coarse regions; pinch points and exit locations are represented using Continuous regions and remainder of the geometry is modelled using Fine nodes. The locations of the various spatial representation types are as shown in Figure 9.4. The section A, depicting the tunnel, is modelled using the Coarse region approach.

Continuous regions were used in different segments of the geometry as follows:
- C1: Section connecting the tunnel and the platform,
- C2 and C3: Sections surrounding internal exits
- C4: Section surrounding the external exit locations.

The remainder of the free space in the geometry was represented using the Fine node approach.

The tunnel was represented using a Coarse region as it was better suited to represent the homogeneous behaviours that would emerge due to the uni-directional motion of the agents in the narrow configuration of the tunnel. Also, the tunnel was located far from the platform and exit locations and therefore the Coarse region could be considered as source nodes to usher the agents into the geometry. Region C1 was modelled as Continuous region as it was better suited to represent the spread of the agents as they emerge from the narrow tunnel and enter the wider platform. The spread of agents is implemented using the Agent Avoidance behaviour which enables the agents to maintain a desired interpersonal space from each other. As depicted in Figure 9.3, an agent P (moving at a velocity V_c) scans the space ahead using a perception box and detects the location of its neighbour P1. The length of the perception box is proportional to the speed of the agent, which ensures that faster moving agents are more responsive in avoiding their neighbours. Once the neighbours have been identified, each agent modifies its trajectory in order to avoid any potential collisions using the resultant steer vector V_r. This behaviour is configured such that neighbouring agents that are further away have lesser influence on the assessing agent. The Agent Avoidance behaviour serves two main purposes; it allows agents to spread out when the space is available and it also allows agents to demonstrate close packing in crowded areas and in constricted regions [8].

Region C2, being an internal exit, was represented as Continuous region to be able to more closely represent the behaviours of the agents at the entrance of the narrower corridor L. Region C3, also being an internal exit, was represented as Continuous region to better model the agents' behaviours as they leave the narrow corridor and move towards the exit locations. Region C4 was modelled as Continuous region as it was connected to two external exits, and was therefore better suited to represent the agent behaviours as they attempt to exit the geometry.

The percentage occupied by the various spatial types was as follows; 73 % of the area was modelled using the Coarse region approach, 23 % of the area was modelled using the Fine node approach and 4 % using the Continuous region. The geometrical setup for the hybrid configuration in bEX-H is shown in Figure 9.5.

A building population of 2,000 agents is initially set up distributed in a 250 m long segment of the tunnel as shown in Figure 9.4. With this distribution of discretisation regions, the agents will traverse three interface transition regions namely; Coarse → Continuous, Continuous → Fine Node, Fine Node → Continuous. The implementation of the transition regions includes behavioural and movement routines that allow agents to transit from spatial type to another [3]. The population characteristics are as shown in Table 9.1. These characteristics have been obtained from sources including [9] and [10]. In this scenario, the response times were assigned to be instantaneous (0 sec) in order to reduce the variation between the different models (Fine and Hybrid).

The buildingEXODUS model has been extensively validated with empirical data in numerous studies [11–15] and was therefore used for the verification of the HSD approach. The Large Tunnel Station Complex case is repeated using only Fine nodes

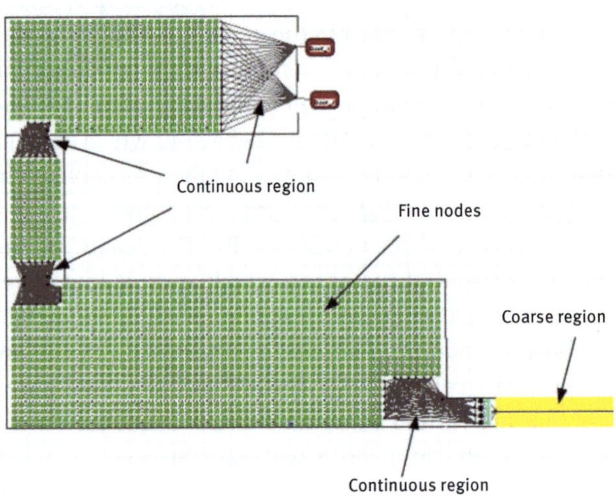

Figure 9.5: Geometrical setup in bEX-H.

Table 9.1: Evacuee profile attribute.

Category	Attribute	Unit	Default value	
			Min	Max
Physical	Mobility	–	1.0	1.0
	Age	Years	17	80
	Weight	Kg	40	90
	Gender	–	Both male and female	
	Agility	–	2	7
	Height	M	1.5	2
Psychological	Response tsime	S	0	0
	Patience	S	1	5
	Drive	-	5	15
Travel speeds	Fast walk	m/s	1.2	1.5
	Walk	m/s	1.08	1.35
	Crawl	m/s	0.24	0.30
	Leap	m/s	0.96	1.20
	Stair-up	m/s	*	*
	Stair-down	m/s	*	*
Experiential	Target exit	–	Nearest exit	
	Familiarity		Nearest exit	
	Itinerary		Null	

such that ten simulation runs were conducted for each case. Each repeat simulation involved the same agents located in the same starting segment of the tunnel; however, their location within that segment was randomised for each repeat.

9.4 Results and discussion

The evacuation time curves for each case are as illustrated in Figure 9.6. These curves represent an average for 10 repeat simulations. In this comparison, the same core software and identical set up was used for both cases, however, differences in the overall times (see Table 9.2) do exist. The Hybrid model predicts an evacuation time which is 3.5 % slower than that predicted by the All-Fine model.

As seen from Figure 9.6, the Hybrid model is predicting faster evacuation times than the All-Fine model up to around 700 sec into the simulation, after which the Hybrid model predicts slower evacuation times. The agents are initially distributed in a segment of the 2 m wide tunnel such that the resultant population density is 4 persons/m^2. The narrow configuration of the tunnel limits the amount of lateral space available to the agents during their movement and therefore reduces the amount of overtaking that can take place. However, there are some significant differences in the representation of the behaviour and movement of the agents through the tunnel in the Hybrid model (Coarse region section) and the All-Fine model.

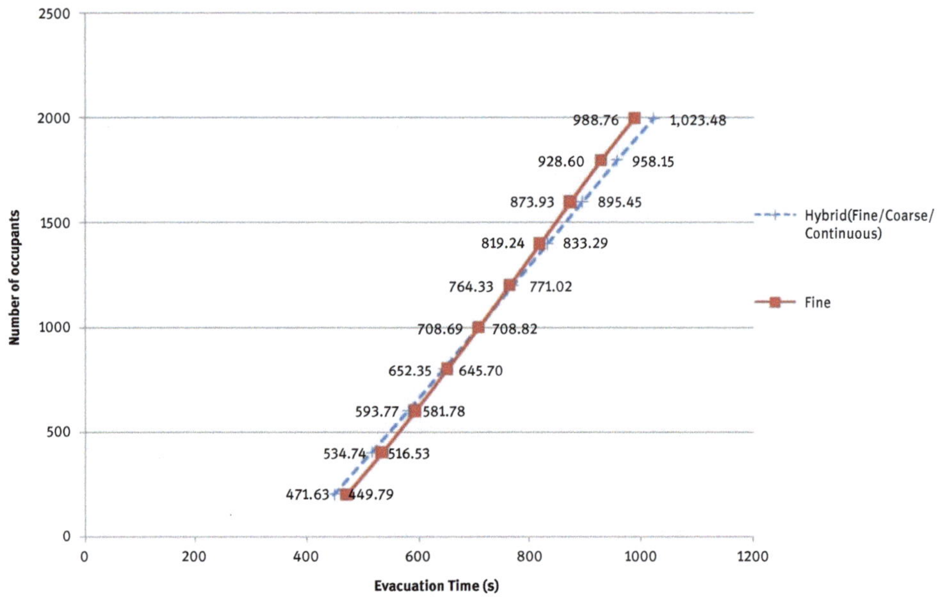

Figure 9.6: Time taken for a specific percentage of people to evacuate using All-Fine and Hybrid configuration.

Table 9.2: Summary of results averaged over 10 simulations.

Spatial representation	Average total evacuation time (s)	Average run time (s)
All-Fine	988.8	80.6
Hybrid	1023.5	44.2

In the All-Fine model, during the initial stage of the simulation, the front part of the stream of agents start to move, while the agents at the back remain stationary until there is enough space available for them to move. In this respect, the population density of the front part of the stream of agents is rather sparse. This results in the flow rate at the exit of the tunnel to be low in the beginning of the simulation and the flow rate gradually increases as the density of the people moving towards the exit of the tunnel builds up. However, in the Hybrid model, due to the lack of agent-agent interactions in the Coarse region approach, the flow rate of agents at the exit of the tunnel is high even at the start of the simulation. This explains the faster evacuation times predicted by the Hybrid approach up to 700 sec into the simulation. In the Hybrid model, flow rate at exit of the tunnel is almost constant throughout the simulation and in the Fine node model, the flow rate increases gradually increases to a peak value which exceeds the peak value in the Hybrid case. After about 700 sec into the simulation, the flow rate at the tunnel exit in the Fine node approach

Table 9.3: Average unit flow rates.

Spatial representation	Average unit flow rate at Door 2 (persons/m/s)
All-Fine	1.55
Hybrid	1.41

starts to exceed the flow rate in the hybrid approach. Moreover, in the Hybrid approach, the Continuous region predict a slower flow rate at the entrance near the external exit location C4 as shown in Table 9.3, which is consistent with the Hybrid model predicting slower evacuation times towards the end of the simulation.

Also shown in Table 9.2 are the average run times for the 10 repeat simulations for each case. It can be seen that the Hybrid model is computationally 45.2 % faster than the All-Fine model. Although the Hybrid model includes the computationally least efficient Continuous approach, however, the speed up gained by the considerably larger proportion of the computationally most efficient Coarse region approach (73 %) compensates for the lower efficiency of Continuous approach. Furthermore, the population distribution in the geometry at about 7 min into the simulation is shown in. The agents in the Continuous regions are seen to exhibit

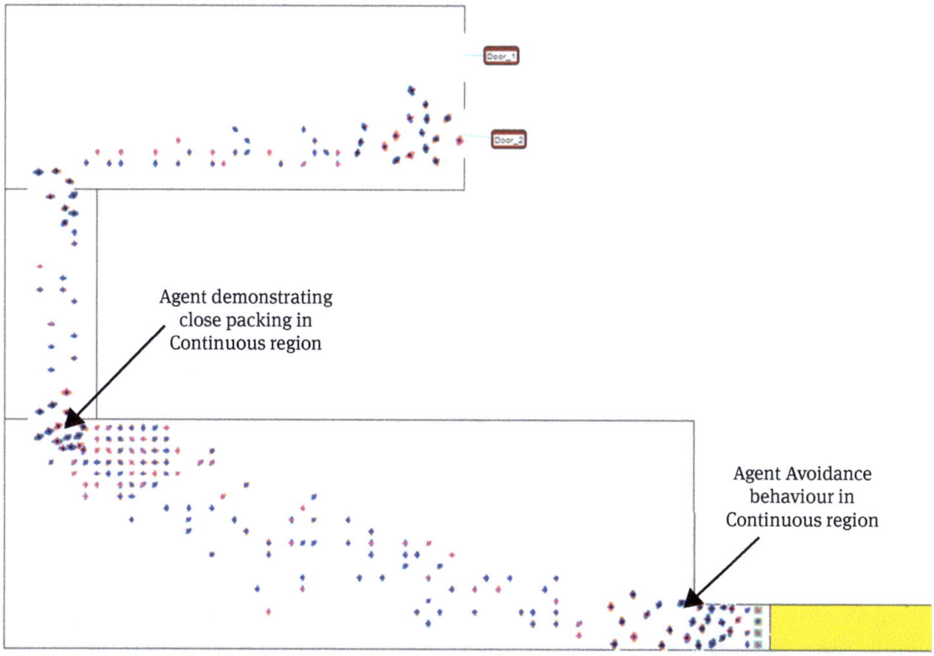

Figure 9.7: Population distribution during a simulation.

Agent Avoidance behaviour which allows agents to dynamically adjust their interpersonal space.

This test case has demonstrated that the HSD approach can be used in an optimal way for performing egress analyses on large complex domains while providing the capability of providing detailed representation of agent behaviours where it is mostly required, for instance, near pinch points, internal and external exits.

9.5 Conclusion

In this paper, we have utilised the HSD approach which allows evacuations simulations to be run using a combination of Coarse regions, Fine nodes and Continuous regions. Each of the spatial representation techniques has their strengths and weaknesses. For instance, if moderately large-scale geometry such as the tunnel station was modelled entirely using Continuous regions, the simulations would provide improved qualitative results and demonstrate emergence of agent behaviours but this configuration would be computationally the least efficient amongst all configurations. Using an all-Fine configuration will not capture the evacuation dynamics at the same level of detail as the all-Continuous as the Fine model represents the person to the nearest nodal location. However, the All-Fine model will provide improved computational performance over the all-Continuous. On the other hand, using an all-Coarse configuration will be computationally very fast and suitable for obtaining very fast preliminary results but it will not provide detailed agent to agent and agent to structure interactions.

However, the strategic combination of all these techniques into the Hybrid model makes it possible to encompass the benefits of all the approaches and provide an optimal environment for representing the movement and behaviour of agents. Moreover, such hybrid configuration enables the accuracy advantages of the Continuous approach to be harnessed but without incurring the large computational overheads of the all-Continuous model. We have demonstrated the benefits of the HSD by applying it to moderately large complex structures such as a tunnel station whereby the model is seen to provide a promising speed up of 45.2 % over the all-Fine approach whilst having the capability of providing detailed agent to agent and agent to structure interactions where it is most required, that is, near internal and external exits and also in regions where there are considerable differences in the geometrical dimensions (for instance a narrow tunnel connected to a wider platform)

Furthermore, the HSD approach involves the transition of agents across the different spatial types. Since the spatial types are of varying granularities, an interface module was developed to regulate the agent flow in the hybrid configuration. Moreover, a Navigational Graph was implemented in the Continuous region to guide the agents towards their goals whilst optimising their goal searching capability. A similar data structure was implemented for the Coarse region approach to be able to represent non-convex Coarse regions. Also, this data structure keeps track of the individual identities of the agents to allow navigation across transition regions.

References

[1] Kuligowski ED, Peacock RD, Hoskins BL. Technical Note 1680. A review of building evacuation models, 2nd ed. U.S. Department of Commerce: Fire Research Division Engineering Laboratory, National Institute Of Standards And Technology, 2010.

[2] Chooramun N, Lawrence PJ, Galea ER. Urban Scale Evacuation Simulation using buildingEXODUS, Interflam 2016,4-6th July 2016, Royal Holloway College, London, UK, Volume 2, pp. 1645-1656 ISBN 978 0 9556548-9-3. 2016.

[3] Chooramun N, Lawrence PJ, Galea ER. An agent based evacuation model utilising hybrid space discretisation. Safety Science. 2012;50(8):1685–1694.

[4] Pretorius M, Gwynne S, Galea ER. Large crowd modelling: An analysis of the Duisburg Love Parade disaster. Fire and Materials. 2013;39(4):301–322.

[5] Galea ER, Lawrence PJ, Gwynne S, Filppidis L, Blackshields D, Cooney D. buildingEXODUS v6.2 theory manual, revision 1.1, FSEG. Greenwich, London: University of Greenwich, 2016.

[6] Buckmann LT, Leather J. Modelling station congestion the PEDROUTE way. Traffic Engineering and Control. 1994;35:373–377.

[7] Craveur JC, Galea ER, Filippidis L. Mesures de vitesses pour les simulations de mouvement de personnes et d'évacuation, Journées du Groupe de Recherche Feux, Cadarche. 2016.

[8] Chooramun N. Thesis (PhD). Implementing a Hybrid Spatial Discretisation within an agent based evacuation model. University of Greenwich, 2011.

[9] Ando K, Ota H, Oki T. Forecasting the flow of people. Railway Research Review. 1988;45:8–14.

[10] Fruin JJ. Pedestrian planning and design. New York: Metropolitan Association of Urban Designers and Environmental Planners Inc, 1971.

[11] Galea ER, Galparsoro JM. A computer based simulation model for the prediction of evacuation from mass transport vehicles. Fire Safety Journal. 1994;22:341–366.

[12] Gwynne S, Galea ER, Lawrence PJ, Filippidis L. Modelling occupant interaction with fire conditions using the buildingEXODUS evacuation model. Fire Safety Journal. 2001;36:327–357.

[13] Galea ER, Wang Z, Veeraswamy A, Jia F, Lawrence PJ, Ewer J. Coupled fire/evacuation analysis of the station nightclub fire. Proceedings of 9th IAFSS Symposium Karlsruhe, Germany, 2008;465–476.

[14] Galea ER, Sharp G, Lawrence PJ. Investigating the representation of merging behavior at the floor stair interface in computer simulations of multi-floor building evacuations. Journal of Fire Protection Engineering. 2008;18:291–316.

[15] Galea ER, Sharp G, Lawrence PJ, Holden R. Approximating the evacuation of the world trade center north tower using computer simulation. Journal of Fire Protection Engineering. 2008;18(2):85–115.

Emildo Marcano

10 DFT study of anthocyanidin and anthocyanin pigments for Dye-Sensitized Solar Cells: Electron injecting from the excited states and adsorption onto TiO_2 (anatase) surface

Abstract: We explored, the absorption spectra, excited states and electronic injection parameters of anthocyanidin and anthocyanin pigments using the level of theory (TD)CAM-B3LYP/6-31+G(d,p). For the most isolated dyes, the distribution pattern of HOMO and LUMO spreads over the whole molecules, which lead an efficient electronic delocalization. The calculated light harvesting efficiencies (LHEs) are all near unity. Methoxy group in peonidin molecule lead the largest oscillator strength and LHE. The presence of water lead a higher spontaneous electronic inject process, with ΔG_{inject} average of −1.14 eV. The ΔG_{inject} order is peonidin < delphinidin < cyanin < cyanidin. Similarly, the adsorption energies (E_{ads}) onto anatase surface model were obtained from level of theory GGA(PBE)/DNP. E_{ads} of anthocyanin–$(TiO_2)_{30}$ complex was calculated to be from 17 to 24 eV, indicating both, the strong interactions between the dyes and the anatase (TiO_2) surface and stronger electronic coupling strengths of the anthocyanin–$(TiO_2)_{30}$ complex, which corresponded to higher observed η. The HOMO and LUMO shape showed the electrons delocalized predominantly on the anthocyanin structure while the LUMO + 1 shape is localized into the $(TiO_2)_{30}$ surface. Therefore, we expected a electronic injection from HOMO to LUMO + 1 in the anthocyanin–$(TiO_2)_{30}$ adsorption complex, after the light absorption.

Keywords: anthocyanin, anatase (TiO_2) surface, adsorption energies, electronic injection

10.1 Introduction

Dye-sensitized solar cells (DSSC) based on organic dyes adsorbed on nanocrystalline TiO_2 electrodes have attracted considerable attention in recent years because of their high incident solar light → electricity conversion efficiency and low cost of production [1, 2]. The driving force in these is the interfacial electron injection from the dye to the semiconductor. Upon absorbing light, the dye molecules are excited from their ground state, which is located energetically in the semiconductor band gap, to an excited state that is resonant with the TiO_2 conduction band. The electron is then transferred to the semiconductor on the ultrafast time scale [3]. The relative yields

https://doi.org/10.1515/9783110467215-010

and rates of electron injection, recombination, and decay of the dye-excited state influence the efficiency of the solar cell [4]. Therefore, improving the efficiency of the solar devices is possible only when the rates and mechanisms of the competing reactions are known and understood. This in turn requires knowledge of the electronic structure of the dyes both, before and after binding to the semiconductor surface [5]. The photochemical properties of different organic sensitizers have extensively been investigated in an attempt to design dyes with maximal visible light absorption coupled to long-lived excited states. However, major effort is still needed in both developing new sensitizers and finding optimal working conditions to improve the photon → current conversion efficiencies [6–8]. In this framework, natural dyes as photosensitizers for DSSCs are very attractive because they are of low cost, abundant in supply and sustainable [9]–[12]. Specifically, several types of natural dyes belonging to anthocyanin have been performed. The anthocyanins belong to the group of natural dyes responsible for several colors in red-blue range, found in fruits, flower and leaves of plants. The dyes extracted from grape, mulberry, blackberry, red Sicilian orange, Sicilian prickly pear, eggplant and radicchio have shown a monochromatic incident photon to current efficiency (IPCE) ranging from 40 % to 69 %. Short circuit photocurrent densities (J_{sc}) up to 8.8 mA/cm^2, open circuit voltage (V_{oc}) ranging from 316 to 419 mV and solar conversion efficiency of 2.06 % [13–15]. Despite several experimental results about the performance of anthocyanin pigments in DSSC, there are no systematic calculations of their electron transference mechanism from the excited states and adsorption onto TiO_2 surface to the best of our knowledge. Within this framework, the main objectives of this paper are to explore, using theoretical calculations, the electronic properties of anthocyanidin and anthocyanin pigments after and before adsorption onto TiO_2 (anatase) surface. The characterization of electronic properties for dyes in gas phase was carried out using the level of theory CAM-B3LYP, employing the 6-31+G(d,p) basis set. The absorption spectra, excited states and electronic injection parameters were obtained and analyzed at TD(CAM-B3LYP)/6-31+G(d,p). Finally, the adsorption energies onto anatase surface model were obtained and analyzed at density functional theory (DFT) level using generalized gradient-corrected approximation (GGA)(Perdew–Burke–Ernzerhof [PBE]) functional and double numerical basis set with polarization (DNP).

10.2 Theory and computational details

The sunlight-to-electricity conversion efficiency (η) of solar cell devices is determined by open-circuit photovoltage (V_{OC}), short-circuit current density (J_{SC}) and the fill factor (FF), as compared to incident solar power (P_{inc}) [1, 16, 17]:

$$\eta = FF \frac{V_{OC} J_{SC}}{P_{inc}} \tag{10.1}$$

J_{SC} is determined by the following equation:

$$J_{SC} = \int LHE(\lambda)\Phi_{inject}\eta_{collect}\, d\lambda \tag{10.2}$$

where $\eta_{collect}$ is the charge collection efficiency, for the same DSSCs differing only in the dye. As is the case for the organic dyes under study, it is reasonable to assume that this parameter is constant. Light harvesting efficiency (LHE (λ) is the fraction of the incident photons that are absorbed by the dye. LHE is related to the oscillator strength (*f*) at a given wavelength. By the following equation, while the larger *f*, the stronger LHE [18]:

$$LHE = 1 - 10^{-f} \tag{10.3}$$

The Φ_{inject} parameter evinces the electron injection efficiency and is related to the driving force ΔG_{inject} of electrons injecting from the excited states of dye molecules to the semiconductor substrate. It can be estimated as [19]:

$$\Delta G_{inject} = E_{OX}^{dye*} - E_{CB}^{TiO_2} \tag{10.4}$$

E_{OX}^{dye*} is the excited state oxidation potential of the dye. $E_{CB}^{TiO_2}$ is the energy conduction band of the TiO_2 semiconductor (−4 eV). E_{OX}^{dye*} can be determined using following formula:

$$E_{OX}^{dye*} = E_{OX}^{dye} - \lambda_{max}^{ICT} \tag{10.5}$$

In eq. 10.5) λ_{max}^{ICT} is the energy of intermolecular charge transfer (ICT).

DFT and time-dependent density functional theory (TDDFT) calculations were performed to determine geometries, electronic structures and electronic absorption spectra of anthocyanin dyes. All the calculations, in gas phase, were performed using GAMESS package [20]. All calculations were performed by employing CAM-B3LYP/6-31+G(d,p). On the basis of eqs 10.3, 10.4, 10.5), we calculated LHE and ΔG_{inject} parameters and analyzed the efficiency of anthocyanin dyes for electron injection from dye's excited state to TiO_2 (anatase) surface [21, 22].

The adsorption of dyes on the anatase cluster was performed with DFT calculations using $DMol^3$ program [23]. The model employed herein to represent the (100) surface of anatase consists of 30 TiO_2 units, terminated with 12 hydrogen atoms, which modeled a TiO_2 nanoparticle [24]. The initial structure has been taken from the crystal of TiO_2 anatase [25]. This model has a diameter of about 1 nm, that has to be compared to nanoparticles of about 2–6 nm used in experiments, and has been used in theoretical study of electronic absorption spectrum of organic compound supported on TiO_2, with application in DSSC [3].

The $(TiO_2)_{30}$ configurations were fully optimized using the GGA. The PBE functional was used to account exchange–correlation effects with DNP basis set. The core electron was treated with DFT-semicore pseudopotentials (DSPPs). After optimization, the adsorption energies (E_{ads}) on the $(TiO_2)_{30}$ cluster were obtained. The latter value was obtained using the equation:

$$E_{ads} = E_{TiO_2} + E_{Anth} - E_{TiO_2+Anth} \tag{10.6}$$

where E_{TiO_2+AmA} is the total energy of anthocyanin–$(TiO_2)_{30}$ complex, E_{TiO_2} is the energy for the anatase surface model and E_{Anth} is the energy for the anthocyanin molecule [26]. Following the above expression, a positive value of E_{ads} indicated a stable adsorption.

10.3 Results and discussion

10.3.1 Geometric optimization and intramolecular charge transferences of anthocyanin dyes

The anthocyanin chromophores molecule (cyaniding, delphinidin, peonidin and cyanin) used to carry out the calculations are displayed in Figure 10.1. They have a positive charge on the molecule, which enables it to absorb light and thus have color [13]. As observed in Figure 10.1, cyanidin and delphinidin differ in the number of hydroxyl groups present in the molecule, while peonidin has a substituted –OH group. Cyanin has a glucoside group. From level of theory CAM-B3LYP/6-31+G(d,p) all structures showed a planar geometric, which facilitates the electronic delocalization in all the structure. The glucoside group in cyanine dye led a small deviation in the torsion angle for the [C3-C2-C1´- C2´] bond (14°).

10.3.2 Frontier molecular orbitals, absorption spectra and LHE

The ground state first singlet excited state excitation process can be mainly assigned to the HOMO–LUMO transition, which correspond to a π-π^* excited singlet state. For the sake of characterizing electronic properties, it is useful to examine the distribution patterns of molecular orbitals [27, 28]. On the other hand, an important thermodynamic requirement of the dyes to be used in DSSC technology is that the HOMO level of the sensitizer has to be sufficiently positive in the redox potential for efficient regeneration of the oxidized dye molecule to its original state by the iodide electrolyte and the LUMO energy of the dye has to be sufficiently higher than the conduction band edge of the semiconductor (E_{CB}). To demonstrate their characteristic electronic structure, the HOMO–LUMO shape from the anthocyanin molecules at CAM-B3LYP/6-31+G(d,p) methods in gas phase, are shown in Figure 10.2, and the energy parameter, including solvent (water) effects with IEFPCM model, are shown in Table 10.1.

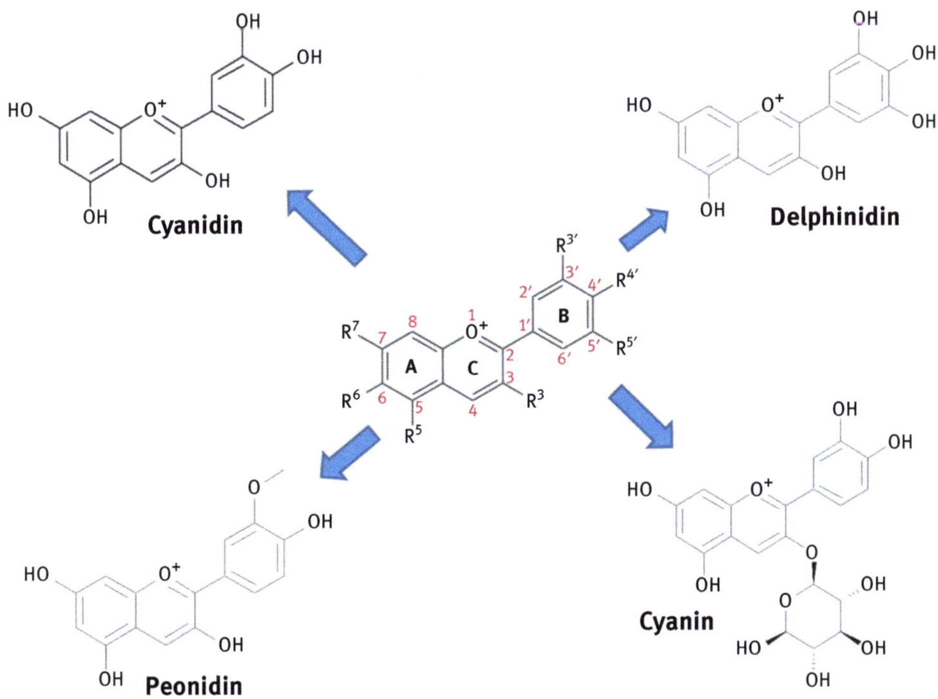

Figure 10.1: Anthocyanin chromophores molecules (cyanidin, delphinidin, peonidin and cyanin) studied in this work.

With exception of cyanin, the distribution pattern of HOMO and LUMO spreads over the whole molecule, as expected, in both, gas phase and water, which lead an efficient electronic delocalization [13]. In fact, the electronic density is shifted from the catechol moiety in ring B to the benzopyran system (ring A and C) (see Figure 10.1). Independently of the substituted groups, in gas phase, the energy values for the HOMO and LUMO orbitals are close to the average −10.46 eV and −5.65 eV, respectively, in agreement with others reports [29], which underestimate the value corresponding to the conduction band of the anatase (−4 eV). When the effects of solvent are account, these values are increased up to −7 eV < −4 eV < −2 eV range [30]. Therefore, the LUMO energy is sufficiently higher than the conduction band edge of TiO_2, and HOMO level is lower than the redox potential of I^-/I_3^- electrolyte to regenerate the oxidized dye (−4,6 eV) [31, 32].

The calculations of the wavelength of maximum absorption (λ_{max}) and others spectroscopic parameter in water are shown in Table 10.2 at TD(CAM-B3LYP)/6-31+G (d,p). Absorption spectra for anthocyanin molecules in gas phase and water are shown in Figure 10.3. The prediction of absorption spectra for structures studied lead two maximum wavelength in 200 nm–250 nm range, and 400 nm–450 nm range, as

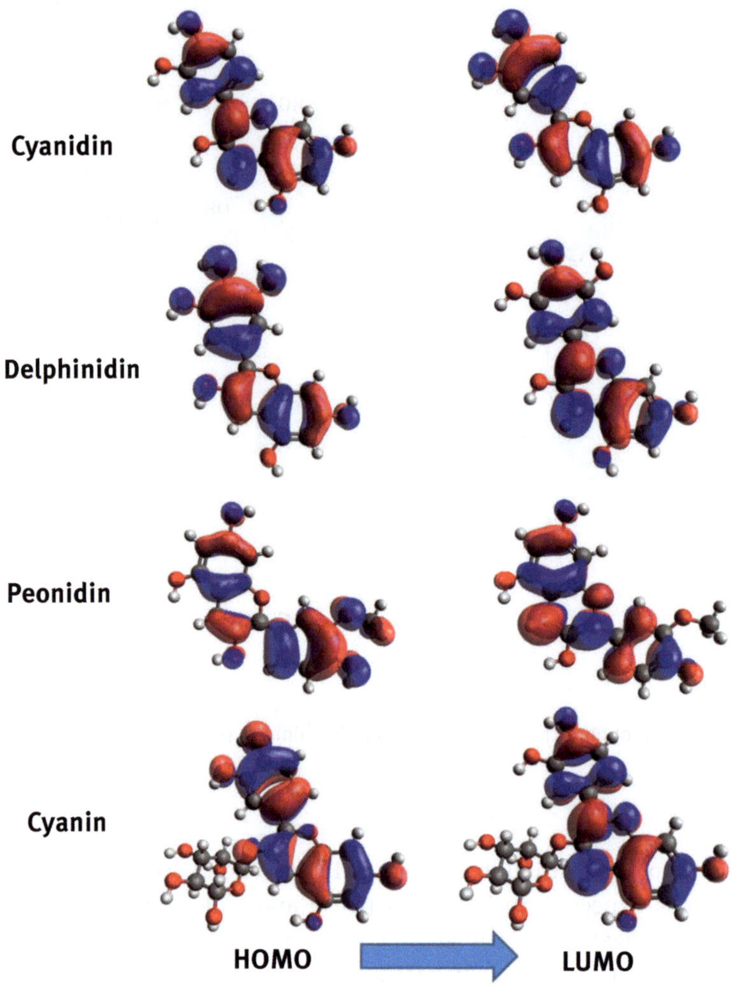

Figure 10.2: HOMO–LUMO shape from the anthocyanin molecules at CAM-B3LYP/6-31+G(d,p) theory level in gas phase.

expected [30, 32, 33]. The λ_{max} values associated with the intramolecular charge transference (ICT), and HOMO → LUMO transition, is in the order cyanidin (446 nm) > peonidin (445 nm) > delphinidin (442 nm) > cyanin (438 nm). From delphinidin molecule, the lack of a hydroxyl group lead a two extra maximum wavelength, associated with the transitions HOMO-4 → LUMO (229 nm) and HOMO → LUMO+2 (205 nm). Similarly, the substitution of a glucoside group in cyanin structure, lead a two extra maximum wavelength, associated with the transitions HOMO → LUMO+1 (250 nm) and HOMO-1 → LUMO+1 (237 nm).

According to the eqs 10.1 and 10.2), LHE is one of the key factors in DSSC. It represents the fraction of the incident photons that are absorbed by the dye. The LHE

Table 10.1: Computed HOMO, LUMO and energy gap (eV) in the gas phase and water at CAM-B3LYP/6-31+G(d,p) level.

	Gas phase			Water		
Molecule	E_{HOMO}	E_{LUMO}	E_{gap}	E_{HOMO}	E_{LUMO}	E_{gap}
Cyanidin	−10.5803	−5.7686	4.8117	−5.6660	−2.6172	5.0489
Delphinidin	−10.4320	−5.6154	4.8166	−5.6405	−2.5796	5.0608
Peonidin	−10.3659	−5.6734	4.6925	−5.6008	−2.6107	4.9901
Cyanin	−10.4992	−5.5578	4.9414	−5.7030	−2.5908	5.1122
Average	−10.4693	−5.6538	4.8156	−5.6526	−2.5996	5.0530

Table 10.2: Wavelength of maximal absorption (λ_{max}/nm), excitation energies (E_e/eV, in water, TD(CAM-B3LYP)/6-31+G(d) level), electronic transition configurations (Assignment), oscillator strengths (*f*) (*f* > 0.1) and LHE for anthocyanin dyes.

Molecule	λ_{max}	Assignment	Ee	*f*	LHE
	446.2	HOMO → LUMO (0.68)	2.78	0.742	0.8187
Cyanidin	229.2	HOMO-4 → LUMO (0.56)	5.41	0.244	
	205.6	HOMO → LUMO+2 (0.49)	6.03	0.283	
	442.2	HOMO → LUMO (0.67)	2.80	0.744	0.8198
Delphinidin	228.4	HOMO-2 → LUMO+1 (0.37)	5.43	0.321	
	206.2	HOMO → LUMO+3 (0.46)	6.01	0.322	
	445.7	HOMO → LUMO (0.66)	2.78	0.767	0.8290
Peonidin	228.4	HOMO-2 → LUMO+1 (0.36)	5.43	0.309	
	207.1	HOMO → LUMO+2 (0.36)	5.99	0.437	
	438.7	HOMO → LUMO (0.68)	2.83	0.723	0.8107
Cyanin	250.8	HOMO → LUMO+1 (0.63)	4.94	0.264	
	237.6	HOMO-1 → LUMO+1 (0.37)	5.22	0.264	

of the dye should be as high as feasible to maximize the photo-current response. As observed in Table 10.2, the calculated LHEs are all near unity. Methoxy group in peonidin molecule lead the largest oscillator strength and LHE. On the other hand, from delphinidin molecule, the lacks of a hydroxyl group lead a decreased in the LHE. Similarly, when a glucoside group is substituted in cyanin structure, a decreased in the LHE value is found.

10.3.3 Free Energy Change of Electron Injection

On the basis of the knowledge of isolated dyes, we extend to study the driving force ΔG_{inject} of electrons injecting from the excited states of anthocyanin dye to the TiO_2 semiconductor substrate to analyze other factors affecting the energy conversion

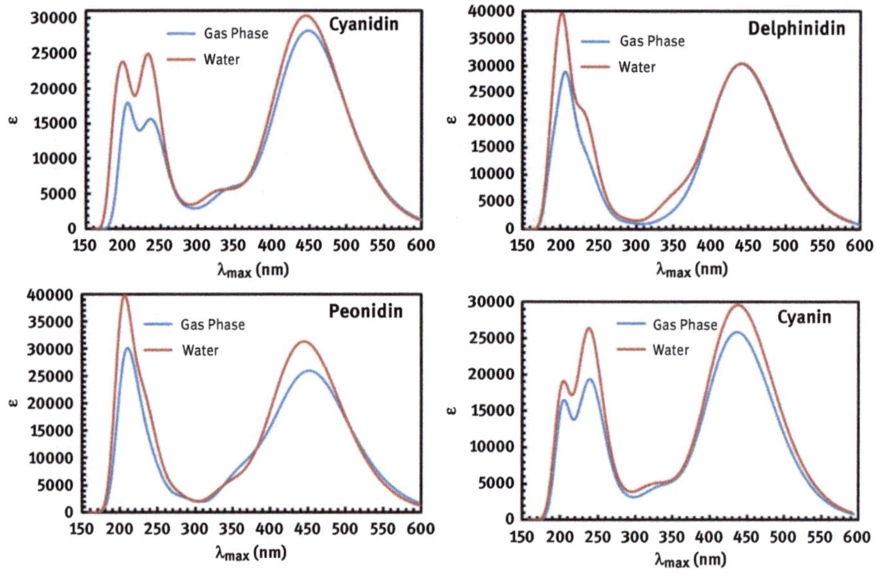

Figure 10.3: Absorption spectra for anthocyanin molecules in gas phase and water at TD(CAM-B3LYP)/6-31+G(d,p) theory level.

efficiency. Therefore, we have used eqs 10.4 and 10.5 to estimate the anthocyanin's excited state oxidation potential and free energy change of electron injection to titanium dioxide TiO_2 surface, in gas phase and water from level of theory TD(CAM-B3LYP)/6-31+G(d,p). E_{OX}^{dye} had been estimated as negative E_{HOMO} [34]. The results are show in Table 10.3. The solvent effects were evidenced in the E_{HOMO} results, which lead a significant decreased in ΔG_{inject}. Therefore, in gas phase, ΔG_{inject} have an average of 3.66 eV, while the water presence lead a higher spontaneous electronic inject process, with ΔG_{inject} average of −1.14 eV. The negative ΔG_{inject} is an indication

Table 10.3: Anthocyanin's excited state oxidation potential and free energy change of electron injection to titanium dioxide TiO_2 surface, in gas phase and water at TD(CAM-B3LYP)/6-31+G(d,p) theory level.

	Gas phase				Water			
Molecule	E^{dye}_{OX}	λ^{ICT}_{max}	E^{dye*}_{OX}	ΔG_{inject}	E^{dye}_{OX}	λ^{ICT}_{max}	E^{dye*}_{OX}	ΔG_{inject}
Cyanidin	10.4320	2.7579	7.6741	3.6741	5.6660	2.7784	2.8876	−1.1124
Delphinidin	10.3659	2.8091	7.5568	3.5568	5.6405	2.8036	2.8369	−1.1631
Peonidin	10.4992	2.7204	7.7788	3.7788	5.6008	2.7820	2.8188	−1.1812
Cyanin	10.4693	2.8316	7.6377	3.6377	5.7030	2.8261	2.8769	−1.1231
Average	10.4416	2.7798	7.6619	3.6619	5.6526	2.7975	2.8551	−1.1449

of spontaneous electron injection from the dye to TiO_2. For the anthocyanin molecules studied, the ΔG_{inject} order is peonidin < delphinidin < cyanin < cyanidin.

10.3.4 Chemisorption on TiO_2-anatase

Tetragonal structure of anatase may be described using two cell edge parameters, *a* and *c*, and one internal parameter, *d* (the length of the Ti–O apical bond) [34]. In this paper, the $(TiO_2)_{30}$ configurations were optimized using the GGA. The results for the geometric parameters described before were *a* = 3.566 Å, *c* = 10.707 Å and *d* = 1.899 Å, which were comparable with experimental values (*a* = 3.782 Å, *c* = 9.502 Å, *d* = 1.979 Å) and others theoretical DFT methodology, where cluster approach methodology had been used [34–36].

In dye-TiO_2 adsorption, the adsorption of dyes through terminal –H atom can be either physisorption (via hydrogen bonding between an oxygen atom on TiO_2 surface and a hydrogen atom of the dye) or chemisorption (an H atom dissociates and the bond is formed between oxygen atoms and the surface titanium atoms of TiO_2). In this paper, we have chosen the second option. The adsorption complex was first fully optimized using the PBE functional together with the double-numerical with polarization performed in the $DMol^3$ program. The optimized structures of anthocyanin–$(TiO_2)_{30}$ adsorption complexes are show in Figure 10.4 and the important optimized bond length and adsorption energy (E_{ads}) are listed in Table 10.4. The bond distances

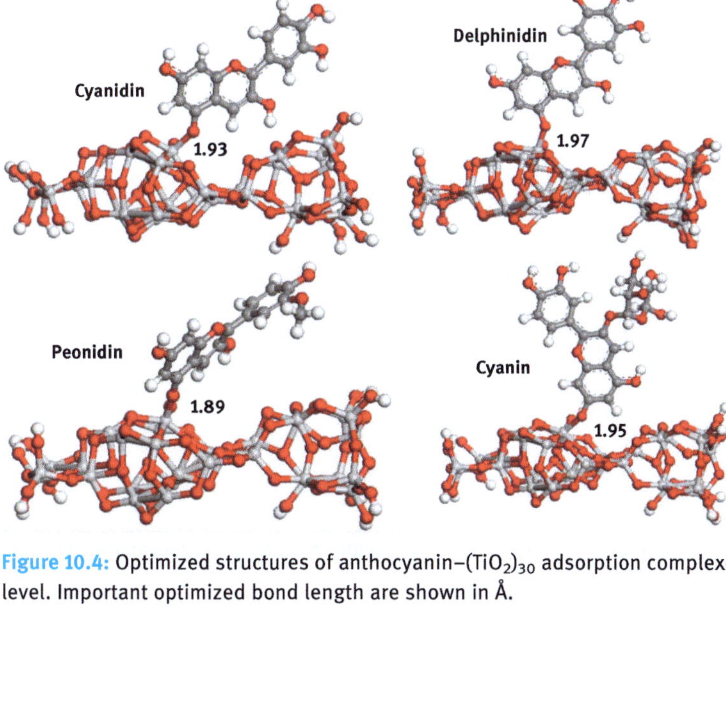

Figure 10.4: Optimized structures of anthocyanin–$(TiO_2)_{30}$ adsorption complexes at PBE/DNP theory level. Important optimized bond length are shown in Å.

Table 10.4: Important optimized bond length Ti-O (Å), adsorption energy (E_{ads}/eV) and sunlight-to-electricity conversion efficiency, η(%) for anthocyanin–$(TiO_2)_{30}$ adsorption complexes.

Molecule	Ti-O	E_{ads}	η (%)
Cyanidin	1.93	17.624	0.37
Delphinidin	1.97	18.749	0.56
Peonidin	1.89	17.889	0.62
Cyanin	1.95	24.107	0.66

between Ti and O atom of dyes were calculated to be in the range of 1.89–1.97 Å. The adsorption energy (E_{ads}) of anthocyanin–$(TiO_2)_{30}$ complex was calculated to be from 17 to 24 eV, indicating the strong interactions between the dyes and the anatase (TiO_2) surface. Table 10.4 shows that a systematic change in the sunlight-to-electricity conversion efficiency (η) was observed as predicted from adsorption energies. Therefore, the higher adsorption energy resulted in the stronger electronic coupling strengths of the anthocyanin–$(TiO_2)_{30}$ complex, which corresponded to higher observed η as expected [37, 38].

In order to explore the possible intramolecular charge transference between anthocyanin dyes and anatase surface, HOMO and LUMO shape were examined by the DFT (PBE) calculations with DNP basis set. Numerical basis set was used because of its reasonable computational cost. Figure 10.5 shows the frontier molecular

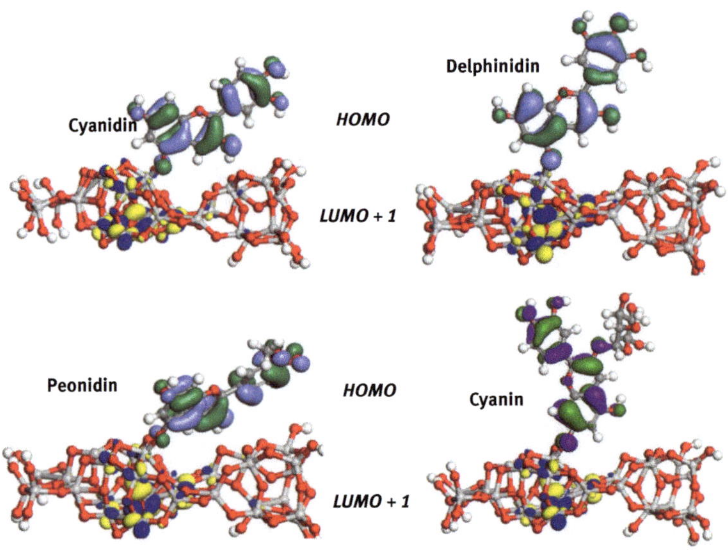

Figure 10.5: The HOMO (blue/green) and LUMO + 1 (blue/yellow) shapes of anthocyanin–$(TiO_2)_{30}$ adsorption complexes at PBE/DNP theory level.

orbitals of anthocyanin–$(TiO_2)_{30}$ adsorption complex in vacuum. The HOMO and LUMO shape showed the electrons delocalized predominantly on the anthocyanin structure while the LUMO + 1 shape is localized into the $(TiO_2)_{30}$ surface. Therefore we expected an electronic injection from HOMO to LUMO + 1 in the anthocyanin–$(TiO_2)_{30}$ adsorption complex, after the light absorption.

Conclusions

We explored the electronic properties of anthocyanidin and anthocyanin pigments after and before adsorption onto TiO_2 (anatase) surface. The characterization of electronic properties for dyes in gas phase and water was carried out using CAM-B3LYP/6-31+G(d,p) methods. The absorption spectra, excited states and electronic injection parameters were obtained and analyzed at TD(CAM-B3LYP)/6-31+G(d,p). The adsorption energies onto anatase surface model were obtained and analyzed at DFT level using GGA(PBE) functional and numerical DNP basis set. For all isolated dyes, the distribution pattern of HOMO and LUMO spreads over the whole molecule, as expected, in both, gas phase and water, which lead an efficient electronic delocalization. The LUMO energy is sufficiently higher than the conduction band edge of TiO_2, and HOMO level is lower than the redox potential of I^-/I_3^- electrolyte to regenerate the oxidized dye. The calculated LHEs are all near unity. Methoxy group in peonidin molecule lead the largest oscillator strength and LHE. The water presence lead a higher spontaneous electronic inject process, with ΔG_{inject} average of −1.14 eV. The negative ΔG_{inject} is an indication of spontaneous electron injection from the dye to TiO_2. For the anthocyanin molecules studied, the ΔG_{inject} order is peonidin < delphinidin < cyanin < cyanidin. The adsorption energy (E_{ads}) of anthocyanin–$(TiO_2)_{30}$ complex was calculated to be from 17 to 24 kcal/mol, indicating the strong interactions between the dyes and the anatase (TiO_2) surface. Therefore, the higher adsorption energy resulted in the stronger electronic coupling strengths of the anthocyanin–$(TiO_2)_{30}$ complex, which corresponded to higher observed η as expected. The HOMO and LUMO shape showed the electrons delocalized predominantly on the anthocyanin structure while the LUMO + 1 shape is localized into the $(TiO_2)_{30}$ surface. Therefore we expected an electronic injection from HOMO to LUMO + 1 in the anthocyanin–$(TiO_2)_{30}$ adsorption complex, after the light absorption.

References

[1] Hagfeldt A, Boschloo G, Sun L, Kloo L, Pettersson H. Dye-sensitized solar cells. Chem Rev. 2010;110:6595–6663.

[2] Vittadini A, Casarin M, Selloni A. Chemistry of and on TiO_2-anatase surfaces by DFT calculations: A partial review. Theor Chem Acc. 2007;117:663–671.

[3] Sánchez De Armas R, Oviedo López J, San Miguel MA. Real-time TD-DFT simulations in dye sensitized solar cells: The electronic absorption spectrum of alizarin supported on TiO_2 nanoclusters. J Chem Theory Comput. 2010;6:2856–2865.
[4] Xie M, Bai FQ, Wang J, Kong CP, Chen J, Zhang HX. Theoretical description of dye regeneration on the TiO_2–dye–electrolyte model. Comput Mater Sci. 2016;111:239–246.
[5] Duncan WR, Prezhdo OV. Electronic structure and spectra of catechol and alizarin in the gas phase and attached to titanium. J Phys Chem B. 2005;109:365–373.
[6] Manzhos S, Giorgi G, Yamashita K. A density functional tight binding study of acetic acid adsorption on crystalline and amorphous surfaces of titania. Molecules. 2015;20:3371–3388.
[7] Jungsuttiwong S, Tarsang R, Surakhot Y, Khunchalee J, Sudyoadsuk T, Promarak V, et al. Theoretical study of α-fluorenyl oligothiophenes as color tunable emissive materials for highly efficient electroluminescent device. Org Electron. 2012;13:1836–1843.
[8] Zhu C, Liang J, Cao Z. Unique metal dicorrole dyes with excellent photoelectronic properties for solar cells: Insight from density functional calculations. J Phys Chem C. 2013;117 13388–13395.
[9] Gao FG, Bard AJ, Kispert LD. Photocurrent generated on a carotenoid-sensitized TiO_2 nanocrystalline mesoporous electrode. J Photochem Photobiol A. 2000;130:49–56.
[10] Hao S, Jihuai W, Huang Y, Lin J. Natural dyes as photosensitizers for dye-sensitized solar cell. Solar Energy. 2006;80:209–214.
[11] Monishka RN. Review: Dye sensitized solar cells based on natural photosensitizers. Renewable Sustainable Energy Rev. 2012;16:208–215.
[12] Hug H, Bader M, Mair P, Glatzel T. Biophotovoltaics: Natural pigments in dye-sensitized solar cells. Appl Energy. 2014;115:216–225.
[13] Calogero G, Yum J-H, Sinopoli A, Di Marco G, Gratzel M, Nazeeruddin MK. Anthocyanins and betalains as light-harvesting pigments for dye-sensitized solar cells. Solar Energy. 2012;86:1563–1575.
[14] Alhamed M, Issa AS, Doubal WA. Studying of natural dyes properties as photo-sensitizer for dye sensitized solar cells (DSSC). J Electron Devices. 2012;16:1370–1383.
[15] Suhaimi S, Shahimin MM, Mohamad IS, Norizan MN. Comparative study of natural anthocyanins compound as photovoltaic sensitizer. Adv Environ Biol. 2013;7:3617–3620.
[16] Feng J, Jiao Y, Ma W, Nazeeruddin MK, GräTzel M, Meng S. First principles design of dye molecules with ullazine donor for dye sensitized solar cells. J Phys Chem C. 2013;117:3772–3778.
[17] Zhang J, Yh K, Hb L, Geng Y, Wu Y, Ya D, et al. Cyano or o-nitrophenyl? Which is the optimal electron-withdrawing group for the acrylic acid acceptor of D-π-A sensitizers in DSSCs? A density functional evaluation. J Mol Model. 2013;19:1597–1604.
[18] O'Regan B, Gratzel M. A low-cost, high-efficiency solar cell based on dye-sensitized colloidal TiO_2 Sfilms. Nature. 1991;353:737–740.
[19] Abdullah MI, Janjua MR, Mahmood A, Ali S, Ali M. Quantum chemical designing of efficient sensitizers for dye sensitized solar cells. Bull Korean Chem Soc. 2013;34:2093–2098.
[20] Schmidt MW, Baldridge KK, Boatz A. GAMESS. J Comput Chem. 1993;14:1347.
[21] Natalia M, Troisi A. Theoretical studies of dye-sensitised solar cells: From electronic structure to elementary processes. Energy Environ Sci. 2011;4:4473–4495.
[22] Zhang J, Kan Y-H, Hai-Bin L, Geng Y, Yong W, Zhong-Min S. How to design proper p -spacer order of the DA dyes for DSSCs? A density functional response. Dyes Pigm. 2012;95:313–321.
[23] Delley B. From molecules to solids with the $DMol^3$ approach. J Chem Phys. 2000;113:7756.
[24] Koch R, Lipton AS, Slawomir F, Renugopalakrishnan V. Arginine interactions with anatase TiO_2 (100) surface and the perturbation of 49Ti NMR chemical shifts–a DFT investigation: Relevance to Renu-Seeram bio solar cell. J Mol Model. 2011;17:1467–1472.
[25] Greeves N. ChemTube3D. http://www.chemtube3d.com/index.html.

[26] Srinivas K, Kumar CR, Reddy MA, Bhanuprakash K, Rao VJ, Giribabu L. A-D organic dyes with carbazole as donor for dye-sensitized solar cells. Synth Met. 2011;161:96–105.
[27] Chang YJ, Chow TJ. Dye-sensitized solar cell utilizing organic dyes containing triarylene conjugates. Tetrahedron. 2009;65:4726.
[28] Kulhánek J, Bureš F. Imidazole as a parent π-conjugated backbone in charge-transfer chromophores. Beilstein J Org Chem. 2012;8:25–49.
[29] El Kouari Y, Migalska-Zalas A, Arof AK, Sahraoui B. Computations of absorption spectra and nonlinear optical properties of molecules based on anthocyanidin structure. Opt Quant Electron. 2015;47:1091–1099.
[30] Eka CP, Yuliarto B, Dipojono HK, Dipojono S. Theoretical investigation of anthocyanidin aglycones as photosensitizers for dye-sensitized TiO_2 solar cells. Adv Mater Res. 2015;1112:317–320.
[31] Duncan RW, Oleg VP. Theoretical studies of photoinduced electron transfer in dye-sensitized TiO_2. Annu Rev Phys Chem. 2007;58:84–143.
[32] Zhang G, Bala H, Cheng Y, Shi D, Lv X, Yu Q, et al. High efficiency and stable dye-sensitized solar cells with an organic chromophore featuring a binary π -conjugated spacer. Chem Commun 2009 2198–2200.
[33] Soto-Rojo R, Baldenebro-López J, Flores-Holguín N, Glossman-Mitnik D. Comparison of several protocols for the computational prediction of the maximum absorption wavelength of chrysanthemum. J Mol Model. 2014;20:2378.
[34] Preat J, Michaux C, Jacquemin D, Perpète EA. Enhanced efficiency of organic dye-sensitized solar cells: Triphenylamine derivatives. J Phys Chem C. 2009;113:16821–16833.
[35] Bourikas K, Kordulis C, Lycourghiotis A. Titanium dioxide (anatase and rutile): Surface chemistry, liquid-solid interface chemistry, and scientific synthesis of supported catalysts. Chem Rev. 2014;114:9754–9823.
[36] Burdett K, Hughbanks T, Miller GJ, Richardson JW, Smith JV. Structural-electronic relationships in inorganic solids: Powder neutron diffraction studies of the rutile and anatase polymorphs of titanium dioxide at 15 and 295 K. J Am Chem Soc. 1987;109:3639–3646.
[37] Henwood D, Carey JD. Molecular physisorption on graphene and carbon nanotubes: A comparative ab initio study. Mol Simul. 2008;34:1019–1023.
[38] Quartarolo AD, Russo N. A computational study (TDDFT and RICC2) of the electronic spectra of pyranoanthocyanins in the gas phase and solution. J Chem Theory Comput. 2011;7:1073–1081 .

Udo Schwingenschlögl, Jiajie Zhu, Tetsuya Morishita, Michelle J.S. Spencer, Paola De Padova, Amanda Generosi, Barbara Paci, Carlo Ottaviani, Claudio Quaresima, Bruno Olivieri, Eric Salomon, Thierry Angot, Guy Le Lay, Harold J.W. Zandvliet and L. C. Lew Yan Voon

11 Elemental Two-Dimensional Materials Beyond Graphene

Abstract: This review article summarizes a few of the papers presented during Symposium II of the International Conference of Pure and Applied Chemistry in July 2016. Properties of monolayers of silicene are first addressed, followed by multilayer silicene and, at the end, a discussion on germanene.

Keywords: silicene, germanene, monolayer, multilayer

11.1 Silicene on substrates

As freestanding silicene has stability issues, understanding the effects of substrates becomes a critical task. Although the material (hexagonal structure) has been deposited successfully on various metallic substrates, including Ag [1], Ir [2], and ZrB_2 [3], strong interaction perturbs the Dirac states and therefore must be avoided. Many semiconducting substrates, such as $MgBr_2$ [4], WSe_2 [5], MoS_2 [6], and solid Ar [7], have been explored theoretically, predicting that the Dirac cone is preserved due to weak van der Waals interlayer interaction.

The lattice mismatches of $MgBr_2$, WSe_2, and MoS_2, all having hexagonal structures, with silicene turn out to be small, see Table 11.1. Ar crystallizes in a cubic structure below 84 K so that the (111) surface is hexagonal and can support silicene. Configurations with different lateral shifts, as shown in Figure 11.1, have similar energies, the maximal differences amounting to 14 meV, 2 meV, 3 meV (high-buckled), and 32 meV (low-buckled) for $MgBr_2$, WSe_2, MoS_2, and solid Ar, respectively, due to flat surface potentials. The interlayer distance is calculated to be more than 3.1 Å (Table 11.1), which reflects weak interaction and corresponds to the fact that the Si buckling height is slightly larger than in the case of freestanding silicene (0.47 Å). It has been reported for MoS_2 that the structure relaxation depends critically on the van der Waals correction method [6, 8]. Both the DFT-D3 and optB86b-vdW methods can reproduce the experimental interlayer distance of 3.0 Å for high-buckled silicene, see Table 11.2. The corresponding values for low-buckled silicene are predicted to be 3.10 Å and 3.15 Å. The vdW-DF1, vdW-DF2, optPBE-vdW, and optB88-vdW methods, on the other hand, underestimate the interlayer interaction.

https://doi.org/10.1515/9783110467215-011

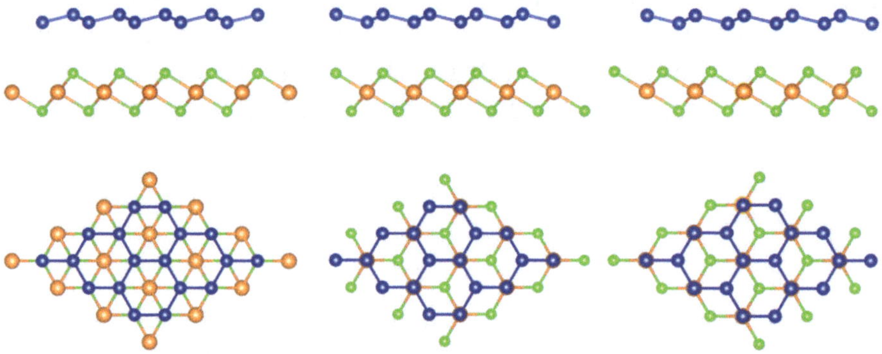

Figure 11.1: Top and side views of silicene on $MgBr_2$ in different configurations. Mg, Br, and Si atoms are shown in brown, green, and blue color, respectively. Reprinted with permission from Zhu, J.; Schwingenschlögl, U. *ACS Appl. Mater. Interfaces* **2014**, *6*, 11675-11681. Copyright 2014 American Chemical Society.

Table 11.1: Structural and electronic parameters of silicene supported by $MgBr_2$, WSe_2, MoS_2, and solid Ar, as calculated by the DFT-D3 method.

	$MgBr_2$ [4]	WSe_2 [5]	MoS_2 [6]	solid Ar [7]
Lattice mismatch (%)	0.3	0.6	3	2
Si buckling height (Å)	0.49	0.51	0.46	0.53
Interlayer distance (Å)	3.23	3.20	3.10	3.10
Band gap (meV)	13	320	20	0

Table 11.2: Structural and electronic properties of silicene on MoS_2, calculated by different van der Waals correction methods.

	DFT-D3	vdW-DF1	vdW-DF2	optPBE-vdW	optB88-vdW	optB86b-vdW
High-buckled						
Si buckling height (Å)	1.91	1.96	2.03	1.94	1.95	1.92
Interlayer distance (Å)	3.01	3.60	3.51	3.31	3.17	3.01
Low-buckled						
Si buckling height (Å)	0.46	0.46	0.48	0.45	0.51	0.48
Interlayer distance (Å)	3.10	3.62	3.52	3.35	3.20	3.15

$MgBr_2$, MoS_2, and solid Ar have been predicted to result in almost gapless Dirac cones (Table 11.1), with the van der Waals correction method influencing the size of the band gap (Table 11.2) [4, 6, 7]. In the case of solid Ar the spin-orbit coupling opens a band gap of 2 meV. In the case of $MgBr_2$ it has been shown that the band gap can be tuned by substitution and intercalation of Li and Na, since at the interface an electric field is induced by charge transfer, which breaks the symmetry of the two silicene sublattices [9]. The band gap is also sensitive to lateral shifts between silicene and substrate, as demonstrated on WSe_2 in Ref. [5], ranging from 0 to 320 meV. Te and Se doping can make the system semiconducting and metallic, respectively. The possibility to achieve large band gaps is promising for ultra-high speed (THz frequency range) field-effect transistors with high on/off current ratio.

11.2 Microscopic mechanism of the oxidation of silicene on Ag(111)

Silicene has the great advantage of easy integration into existing circuitry that is already based on Si technology. The application of silicene to nanoscale devices is, however, currently hindered by the existence of unsaturated (dangling) bonds on its surface, which makes it highly reactive under atmospheric conditions. It is, however, crucial to be able to control the stability of silicene under atmospheric conditions in order to fabricate silicene-based nanodevices. In particular, elucidating the reaction process of silicene with oxygen is urgently required. We here review recent ab initio molecular-dynamics (AIMD) simulations to uncover the atomistic mechanism of the oxidation process of the silicene overlayer on the Ag(111) surface [10].

The AIMD simulations were performed for the 3×3 honeycomb silicene lattice on the 4×4 Ag(111) surface within the framework of density functional theory as implemented in the Vienna Ab Initio Simulation Package. The exchange-correlation functional in the Perdew–Burke–Ernzerhof form was used and the ion-electron interaction was described by the projector augmented wave method. The electronic wave functions were expanded in a plane-wave basis set with an energy cutoff of 400 eV. The Ag substrate consists of five atomic layers with the bottom layer fixed.

It was found that there exist barrier-less oxygen chemisorption pathways around the outer Si atoms of the silicene overlayer, indicating that oxygen can easily react with a Si atom to form an Si-O-Si configuration, once the molecule finds an entrance to the pathway on the rugged energy landscape provided by the silicene overlayer. The Si-O bond formed in the reaction of oxygen is not covalent but rather ionic, which results from the charge transfer from the Si atom to the O atom. It was found that about $0.8|e|$ is transferred from Si to O atoms. As a result, the nature of the intermediate sp^2/sp^3 bonding in the silicene overlayer is substantially degraded upon oxidation.

In the reaction process involving multiple O_2 molecules, a synergistic effect between the molecular dissociation and subsequent structural rearrangements was found to accelerate the oxidation process, especially at a high oxygen dose. This effect enhances self-organized formation of sp^3-like tetrahedral configurations (consisting of Si and O atoms), which results in collapse of the two-dimensional silicene structure and its exfoliation from the substrate (see Figure 11.2) [10].

It was also found that the electronic properties of silicene can be significantly altered by oxidation. Figure 11.2 shows the atom-resolved density of states (DOS) for a

Figure 11.2: Decomposed electronic DOS for (a) a threefold-coordinated Si atom and (b) a fourfold-coordinated Si atom in the oxidized silicene (as indicated by the bold arrows and white circles). The zero of energy in the DOS plots is aligned to the Fermi energy. The red, yellow, and pink balls indicate O, Si, and Ag atoms, respectively [10].

Si atom having three neighboring Si atoms, as in the silicene honeycomb lattice. It is clear that the electronic bands near the Fermi energy have a high intensity and are dominated by the p_z electrons from the dangling bond on the Si atom, showing a metallic nature. In contrast, the atom-resolved DOS for a four-coordinated Si atom (with two O atoms and two Si atoms), having a highly tetrahedral configuration [Figure 11.2], shows completely different characteristics with the electronic bands near the Fermi energy being substantially reduced, especially those from the p_z electrons, due to capping of the dangling bonds with O atoms. We thus conclude that the metallic nature of silicene is reduced as oxidation proceeds. This tendency has also been observed in a recent experimental study [11].

11.3 Multilayer silicene

Recent work on multilayer silicene using an energy-dispersive grazing incidence x-ray diffraction (ED-GIXRD) study is now summarized [12]. The growth of multilayer silicene has been realized at a temperature of about 200 °C on top of the initial archetype 3×3 monolayer silicene phase on a single crystal Ag(111) surface. It proceeds in successive flat terraces separated by 0.3 nm. These terraces show a unique √3×√3 reconstruction whose cell size was found to be 6.477 ± 0.015 Å by ED-GIXRD, in agreement with scanning tunneling microscopy measurements. Figure 11.3 displays the ED-GIXRD pattern from 10 monolayer (ML) thick multilayer silicene.

The first- and second-order in-plane reflections at $q_{xy} = 0.970 \pm 0.005$ Å^{-1} and $q_{xy} = 1.939 \pm 0.005$ Å^{-1}, corresponding to the √3×√3 cell size of multilayer silicene, $a_{ML} = 6.477 \pm 0.015$ Å, and, in addition, the out-of-plane reflection at $q_z = 2.033 \pm 0.005$ Å^{-1}, corresponding to $d_{zML} = 3.090 \pm 0.010$ Å, as previously reported, are clearly recorded. Particularly noteworthy is that absolutely no Si(220) reflection was detected (see inset on the right side of Figure 11.3). This proves that the whole body of the multilayer

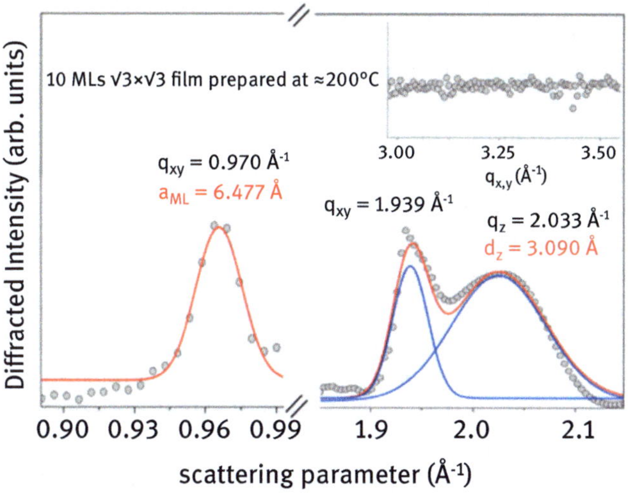

Figure 11.3: ED-GIXRD pattern collected from a √3×√3 multilayer film (10 MLs) grown on Ag(111) at 200 °C and Gaussian fit (red line) of each reflection. First-order in-plane q_{xy} = 0.970 Å^{-1} ($FWHM_{xy}$ = 0.0467 Å^{-1}) and out-of-plane q_z = 2.033 Å^{-1} ($FWHM_z$ = 0.2090 Å^{-1}) reflections; second-order in-plane reflection q_{xy} = 1.939 Å^{-1} ($FWHM_{xy}$ = 0.0790 Å^{-1}) (blue line). The inset displays the ED-GIXRD pattern around 3.27 Å^{-1}; neither a Si(220) peak nor its relaxation are observed. The positions of the peaks are the centroid of the Gaussians. Adapted from Figure 11.3 of Ref. [12].

silicene film possesses the a_{ML} = 6.477 ± 0.015 Å in-plane lattice parameter, totally different from: i) a bulk-like Si(111) arrangement terminated by a √3×√3-Ag reconstruction ($a_{Si(111)\sqrt{3}}$ = 6.655 ± 0.015 Å), ii) a tetragonally strained bulk-like Si(111) arrangement terminated by a Si(111)-√3×√3-Ag reconstruction; and finally iii) a bulk-like Si(111) arrangement terminated by a Si(111)-√3×√3 intrinsic reconstruction. These measurements [12] have demonstrated the existence of multilayer silicene in the low temperature growth regime.

11.4 Germanene

Germanene, the germanium analogue of graphene, has been successfully synthesized in 2014 by three different groups: in July 2014 by the Gao group [13] and in September 2014 by the Le Lay [14] and Zandvliet [15] groups.

The structural and electronic properties of germanene and silicene are predicted to be very similar to graphene [16, 17]. However, there are also a few distinct differences, such as the honeycomb lattice, which is fully planar for graphene, but buckled for silicene and germanene [18]. Despite this buckling, tight binding and density functional theory calculations have revealed that the Dirac properties of silicene and germanene are not destroyed. Near the K and K′ points of the Brillouin zone the energy bands of silicene and germanene are predicted to be linear in k. Besides this buckling of the honeycomb lattice, there is another very salient difference between graphene and silicene/germanene, namely the size of the spin-orbit gap. The spin-orbit gap in graphene, silicene, and germanene makes that these materials are not true semimetals but rather two-dimensional topological insulators. The spin-orbit gap in graphene is only 24 μeV [19], whereas the spin-orbit gaps in silicene and germanene are substantially larger, namely 1.55 meV and 23.9 meV [20], respectively. Silicene and particularly germanene are thus ideal candidates to exhibit the quantum spin Hall effect at experimentally accessible temperatures.

In Figure 11.4 a scanning tunneling microscopy image of germanene recorded by Bampoulis et al. [15] is shown. The germanene layer was synthesized on a Ge_2Pt crystal. Bampoulis et al. [15] managed to resolve the atomic structure of the unit cell of germanene. As can be seen in Figure 11.4, the honeycomb cell of germanene is buckled, i. e., half of the atoms reside in a 'high' position, whereas the other half of the atoms reside in a 'low' position. The buckling is only 0.2 Å, i. e., substantially smaller than the buckling predicted for freestanding germanene (0.65 Å) [17].

So far germanene has been grown only on metallic substrates [13–15, 18, 21–24]. Unfortunately, metallic substrates are often detrimental for the Dirac nature of the two-dimensional materials, because the relevant electronic states near the Fermi

Figure 11.4: Scanning tunneling microscopy image recorded at a sample bias of 0.5 V and a tunnel current of 0.2 nA. The image size is 4.5 nm × 4.5 nm. The nearest neighbor distance between the Ge atoms is 2.5 ± 0.1 Å and the buckling is 0.2 Å. Data taken from Ref. [15].

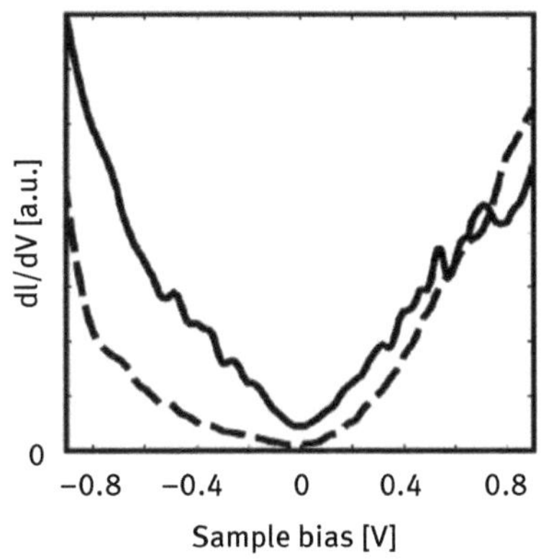

Figure 11.5: Differential conductivities of a germanene layer synthesized on a molybdenum disulfide substrate (solid line) and of bare molybdenum disulfide (dashed line). The set points are −1.4 V and 0.6 nA. Data from Ref. [23].

level can hybridize with the electronic states of the metallic substrate. In 2016, Zhang et al. [23] succeeded to grow germanene on molybdenum disulfide, a band gap material. They performed scanning tunneling spectroscopy measurements and showed that the DOS of germanene exhibits a well-defined V-shape, which is one of the hallmarks of a two-dimensional Dirac system (see Figure 11.5). The Dirac point is located very close to the Fermi level, but the DOS does not completely vanish at the Dirac point.

Density functional theory calculations showed that this non-zero DOS at the Fermi level is due to the fact that the buckling of the germanene layer is somewhat larger than the buckling of freestanding germanene. For a buckling larger than about 0.8 Å two electronic states, that originate from two parabolic bands of germanene, cross the Fermi level at the Γ point of the surface Brillouin zone. The exact effect of this non-zero DOS on the Dirac properties remains to be investigated. In any case density functional theory calculations by Amlaki et al. [24] have revealed that the topological insulator character of germanene remains intact upon the interaction with molybdenum disulfide.

11.5 Summary

We have elaborated on a few properties of monolayer silicene, multilayer silicene, and germanene. It has been shown that the Dirac cone of silicene is preserved on some substrates, whereas on other substrates energy gaps are created. Silicene oxidation studies show that this process can lead to exfoliation and significant changes of the electronic properties of silicene. Energy-dispersive grazing incidence x-ray diffraction has been used to confirm the growth of multilayer silicene in the low temperature regime. Finally, recent work on the synthesis of germanene on metallic as well as semiconducting substrates has been briefly discussed. High-resolution scanning tunneling microscopy data reveal that germanene exhibits a buckled honeycomb lattice. The DOS of germanene synthesized on various substrates has a V-shape, which strongly hints to a two-dimensional Dirac material.

References

[1] Vogt P, De Padova P, Quaresima C, Avila J, Frantzeskakis E, Asensio MC, et al. Silicene: Compelling experimental evidence for graphenelike two-dimensional silicon. Phys Rev Lett. 2012;108:155501.

[2] Meng L, Wang Y, Zhang L, Du S, Wu R, Li L, et al. Buckled silicene formation on Ir(111). Nano Lett. 2013;13:685–690.

[3] Fleurence A, Friedlein R, Ozaki T, Kawai H, Wang Y, Yamada-Takamura Y. Experimental evidence for epitaxial silicene on diboride thin films. Phys Rev Lett. 2012;108:245501.

[4] Zhu J, Schwingenschlögl U. Structural and electronic properties of silicene on MgX_2 (X = Cl, Br, and I). ACS Appl Mater Interfaces. 2014;6:11675–11681.

[5] Zhu J, Schwingenschlögl U. Stability and electronic properties of silicene on WSe_2. Mater Chem C. 2015;3:3946–3953.

[6] Zhu J, Schwingenschlögl U. Silicene on MoS_2: Role of the van der Waals interaction. 2D Mater. 2015;2:045004.

[7] Sattar S, Hoffmann R, Schwingenschlögl U. Solid argon as a possible substrate for quasi-freestanding silicene. New J Phys. 2014;16:065001.

[8] Chiappe D, Scalise E, Cinquanta E, Grazianetti C, Van Den Broek B, Fanciulli M, et al. Two-dimensional Si nanosheets with local hexagonal structure on a MoS_2 surface. Adv Mater. 2014;26:2096–2101.

[9] Zhu J, Schwingenschlögl U. Band gap opening in silicene on $MgBr_2$(0001) induced by Li and Na. ACS Appl Mater Interfaces. 2014;6:19242–19245.

[10] Morishita T, Spencer MJ. How silicene on Ag(111) oxidizes: Microscopic mechanism of the reaction of O_2 with silicene. Sci Rep. 2015;5:17570.

[11] Du Y, Zhuang J, Liu H, Xu X, Eilers S, Wu K, et al. Tuning the band gap in silicene by oxidation. ACS Nano. 2014;8:10019–10025.

[12] De Padova P, Generosi A, Paci B, Ottaviani C, Quaresima C, Olivieri B, et al. Multilayer silicene: Clear evidence. 2D Mater. 2016;3:031011.

[13] Li L, Lu S-Z, Pan J, Qin Z, Wang Y-Q, Wang Y, et al. Buckled germanene formation on Pt(111). Adv Mat. 2014;26:4820–4824.

[14] Dávila ME, Xian L, Cahangirov S, Rubio A, Le Lay G. Germanene: A novel two-dimensional germanium allotrope akin to graphene and silicene. New J Phys. 2014;16:095002.

[15] Bampoulis P, Zhang L, Safaei A, Van Gastel R, Poelsema B, Zandvliet HJW. Germanene termination of Ge_2Pt crystals on Ge(110). J Phys Cond Mat. 2014;26:442001.

[16] Guzmán-Verri GG, Lew Yan Voon LC. Electronic structure of silicon-based nanostructures. Phys Rev B. 2007;76:075131.

[17] Cahangirov S, Topsakal M, Aktürk E, Şahin H, Ciraci S. Two- and one-dimensional honeycomb structures of silicon and germanium. Phys Rev Lett. 2009;102:236804.

[18] Acun A, Zhang L, Bampoulis P, Farmanbar M, Van Houselt A, Rudenko AN, et al. Germanene: The germanium analogue of graphene. J Phys Cond Matt. 2015;27:443002.

[19] Yao Y, Ye F, Qi X-L, Zhang S-C, Fang Z. Spin-orbit gap of graphene: First-principles calculations. Phys Rev B. 2007;75:041401.

[20] Liu CC, Feng W, Yao Y. Quantum spin hall effect in silicene and two-dimensional germanium. Phys Rev Lett. 2011;107:076802.

[21] Zhang L, Bampoulis P, Van Houselt A, Zandvliet HJW. Two-dimensional Dirac signature of germanene. Appl Phys Lett. 2015;107:111605.

[22] Dávila ME, Le Lay G. Few layer epitaxial germanene: A novel two-dimensional Dirac material. Sci Rep. 2016;6:20714.

[23] Zhang L, Bampoulis P, Rudenko AN, Yao Q, Van Houselt A, Poelsema B, et al. Structural and electronic properties of germanene on MoS_2. Phys Rev Lett. 2016;116:256804.

[24] Amlaki T, Bokdam M, Kelly PJ. Z_2 invariance of germanene on MoS_2 from first principles. Phys Rev Lett. 2016;116:256805.

Index

https://doi.org/10.1515/9783110467215-012

Bei Fragen zur Produktsicherheit wenden Sie sich bitte an:
If you have any questions regarding product safety,
please contact:

Walter de Gruyter GmbH
Genthiner Straße 13
10785 Berlin
productsafety@degruyterbrill.com